Martin Wagner
The Narratology of Observation

Paradigms

Literature and the Human Sciences

Edited by
Rüdiger Campe · Paul Fleming

Volume 7

Martin Wagner

The Narratology of Observation

Studies in a Technique of European Literary Realism

DE GRUYTER

ISBN 978-3-11-059518-5
e-ISBN (PDF) 978-3-11-059434-8
e-ISBN (EPUB) 978-3-11-059359-4
ISSN 2195-2205

Library of Congress Control Number: 2018947262

Bibliographic information published by the Deutsche Nationalbibliothek
The Deutsche Nationalbibliothek lists this publication in the Deutsche Nationalbibliografie;
detailed bibliographic data are available on the Internet at http://dnb.dnb.de.

www.degruyter.com

Acknowledgments

Simple as the argument advanced in this book is, I could not have formulated it without the critique and guidance of Rüdiger Campe, Paul Fleming, Jennie Han, Stéphane Lojkine, Paul North, Kirk Wetters, and the two anonymous reviewers of my manuscript. I would also like to acknowledge Yale University, Yonsei University, and the University of Calgary for providing me with the financial support and intellectual environment that helped me write this book. Richard Slipp worked with me on the preparation of the final manuscript. I dedicate this book to Franziska.

https://doi.org/10.1515/9783110594348-005

Contents

Table of Figures

https://doi.org/10.1515/9783110594348-009

Introduction

Literary observation

This book is devoted to a narrative sequence that we have all already encountered in our readings: more or less suddenly, a person (or object) appears and is described in some detail, and then this person (or object) is set in motion so that a narrative begins, however minimal this narrative may be. For reasons that I will get to later, I call this sequence, borrowing from the history of science, an *observation* (and I will sometimes speak, more precisely, of *literary* observation, to be able to distinguish it in the pages of this book from scientific observation). Literary observations, I argue, perform for the reader a process of perception that moves from the more or less sudden appearance of a single image of a person (or object) to the sustained watching in which this image is set in motion. They mimic the way in which we become aware of new aspects of reality and then see them develop over time.

Literary observations can be explicitly marked as a process of visual perception by showing the image and its movement through the eyes of a character in the story (or, as a special case of this, through the eyes of a homodiegetic narrator). But literary observations can also just implicitly imitate a process of perception (this is the case whenever a heterodiegetic narrator proceeds from the more or less sudden introduction of a static description to a process of narration in which this static image is set in motion).

The function or effect of these performed processes of perception is something that I want to call, borrowing very loosely from Roland Barthes, a "reality effect." Through the observations, the reader is presented with a world that appears visual and 'real' not just in the still lifes of description, but also in its dynamic development over time. Readers are invited to experience the text as if they were visually exploring the world.

In analyzing the procedure of literary observation through readings from the early eighteenth to the late nineteenth centuries, I construct in this book a narratology of observation that aims to enhance our understanding of literary realism in both its technical, stylistic features and its cultural preconditions.[1] With

---

1 I use the term realism here to denote two slightly separate things: the stylistic features typical of the period of literary realism (of the nineteenth century and its precursors in the eighteenth century) and, more generally, a form of literature whose concern it is to evoke reality. Using the term realism in these two senses is somewhat imprecise, perhaps (as at least Roman Jakobson famously argued [Jakobson 1987, 20–21]). However, this might be an unavoidable imprecision,

https://doi.org/10.1515/9783110594348-011

respect to the stylistics of literary realism, the present study contributes to our understanding of the relation between description and narration, which has been at the center of many debates on literary realism. Literary observation is a way for writers to carry the visuality that is conventionally associated with description over into the narration and thus to bridge the gap between description and narration in a text. But the narratology that I present is also very much a *cultural* narratology. Over the course of this book, I explore a range of cultural discourses and techniques – from classical poetics and art criticism through psychology, morphology, and evolutionary biology to the social surveillance in the emerging big cities of the nineteenth century – that either enable or inhibit the seemingly so simple shift from the description of a static image to the narration of a sequence of events.

An example

Before I develop any of these grander claims about the mechanics and cultural preconditions of literary observation in more detail, I want to point to one example. The opening paragraphs of Gustave Flaubert's novel *Madame Bovary* (1857) contain, I argue, a typical procedure of literary observation, while also already raising our awareness for some possible complications in this procedure (and we will see many more complications as we proceed). The passage is narrated from the perspective of a schoolboy sitting in class and watching the arrival of a new pupil:

> Nous étions à l'Étude, quand le Proviseur entra, suivi d'un *nouveau* habillé en bourgeois et d'un garçon de classe qui portait un grand pupitre. Ceux qui dormaient se réveillèrent, et chacun se leva comme surpris dans son travail.
> Le Proviseur nous fit signe de nous rasseoir, puis, se tournant vers le maître d'études :
> – Monsieur Roger, lui dit-il à demi-voix, voici un élève que je vous recommande, il entre en cinquième. Si son travail et sa conduite sont méritoires, il passera *dans les grands*, où l'appelle son âge.
> Resté dans l'angle, derrière la porte, si bien qu'on l'apercevait à peine, le *nouveau* était un gars de la campagne, d'une quinzaine d'années environ, et plus haut de taille qu'aucun de nous tous. Il avait les cheveux coupés droit sur le front, comme un chantre de village, l'air raisonnable et fort embarrassé. Quoiqu'il ne fût pas large des épaules, son habit-veste de drap vert à boutons noirs devait le gêner aux entournures et laissait voir, par la fente des parements, des poignets rouges habitués à être nus. Ses jambes, en bas bleus, sortaient d'un pantalon jaunâtre très tiré par les bretelles. Il était chaussé de souliers forts, mal cirés,

---

as the period of literary realism itself is evidently defined and judged by a concept of realism that transcends this period's historical limits.

garnis de clous.

On commença la récitation des leçons. Il les écouta de toutes ses oreilles, attentif comme au sermon, n'osant même croiser les cuisses, ni s'appuyer sur le coude, et, à deux heures, quand la cloche sonna, le maître d'études fut obligé de l'avertir, pour qu'il se mît avec nous dans les rangs. (Flaubert 1998, 11–12)

We were at prep when the Headmaster came in, followed by a 'new boy' not wearing school uniform, and by a school servant carrying a large desk. Those who had been asleep woke up, and we all rose to our feet as though we had been interrupted at our work.

The Headmaster motioned to us to be seated; then, turning to the master on duty:

'Monsieur Roger,' he said in a low voice, 'this is a pupil I'm putting in your hands. He's starting in the fifth. If his work and his conduct warrant it, he'll be moved up to the "seniors", which is where he should be, given his age.'

Still standing well back, in the corner behind the door, so that he was almost invisible, the 'new boy' was a country lad of about fifteen who towered over the rest of us. He wore his hair in a straight fringe across his brow, like a village choirboy, and he looked sensible and very ill at ease. Although his shoulders were not particularly broad, his green cloth jacket with black buttons seemed tight around the armholes, and revealed, through his slits of his cuffs, sun-reddened wrists unaccustomed to being covered. His legs, clad in blue stockings, stuck out below a pair of yellowish trousers that were hitched up very high by his braces. His shoes were heavy, badly shined, and studded with nails.

We began reciting our lessons. He listened attentively, concentrating as though listening to a sermon, not daring even to cross his legs or lean on his elbow, and, at two o'clock, when the ball rang, the master had to tell him to line up with us all. (Flaubert 2004a, 5)

This scene follows the pattern of observation that I outlined above almost self-consciously. Suddenly, the door of the classroom opens, waking the students from their slumber to see someone new ("un *nouveau*") enter. We receive a brief first description of the new boy: he appears "en bourgeois," apparently not wearing the obligatory school uniform and thus standing out from his environment. The text then repeats its initial gesture of presenting a curious new sight by making explicit what just happened. The school's principal (denoted in the French original quite tellingly with the term *Proviseur:* literary, someone who sees before, or foresees [*prévoir*]) tells us what there is to see: "voici un élève que je vous recommande" [this is the pupil I'm putting in your hands]. The narrator then follows up on this introduction of the new student whom we are made to see (*voici*) with a detailed description.

What we then read is in some sense a classic description, following rules already set out by the Greek rhetorician Aphthonius in the fourth century CE. As Aphthonius prescribes, one should proceed in the description of a person from head to toe (Hamon 1991, 24), and this is precisely what we see in Flaubert's novel. Even the emphasis on particulars in the passage from Flaubert follows classical models. As Beth Innocenti argues in her reconstruction of a Roman poetics of "vivid description," the mentioning of particular features was understood to be a key technique to allow listeners to visualize the described persons and objects (Innocenti 1994, 370). However, we probably also recognize that the kind of particularities that we are made to see here are a typical feature of nineteenth-century realism, with its attention to the imperfect. We learn about the ill-fitting, dirty clothes and the strained body (just consider the red wrists, sticking out from underneath the tight-fitting school jacket). As Jonathan Culler reminds us in an article on the realism in *Madame Bovary*, "[i]n the mid-nineteenth century realism has especially the connotation of reference to the low or vulgar that was not previously accommodated by literary discourse" (Culler 2007, 689).

However, what is more crucial here for my purposes than the new attention to the ugly aspects of social reality and to human imperfection is that the description is not left there as such, as a static image. We do not simply pass on from some background description back to the narrative. Think here, for the sake of contrast, of scenes in which a room is being described upon entering it, or in which a landscape is being described that the protagonist passes through. In such scenes, which are abundant in the realist tradition as well, the text simply moves from the description back to the narrative sequence of actions: the described things are left in their place, so to speak, and the text's attention focuses once more on the actions. There may be various functions that such background descriptions fulfill: as Zola argued, for instance, these descriptions are crucial for a science-like literature that explains how specific characters emerge from a particular milieu (like plants and animals that thrive only in a certain habitat; see Hamon 1991, 155 – 162). These background descriptions can symbolize, moreover, personal traits of a character, or they can stage how a character sees the world. Tove Holmes, for example, has shown that much of Adalbert Stifter's *Der Nachsommer* [Indian Summer] (1857) proceeds through a series of descriptions that reflect the steady inner growth of the protagonist. The descriptions, Holmes argues, track the protagonist's increasingly complex 'view' of the world (Holmes 2010). Alternatively, these descriptions can, as Barthes famously claimed in his essay on the "effet de réel" [reality effect], provide some information that, while useless for the advancement of the plot or the characterization of the characters, points to the category of the real as such. Finally, and in addition to all the other functions just listed, such background descriptions can provide us with a visual image of the world, presenting the world as if it were there, before our eyes.[2]

But the observation with which Gustave Flaubert opens his debut novel *Madame Bovary* is different from these background descriptions. Here, the narra-

---

2 The functions of description are analyzed in more detail in chapter 1.

tor's attention remains on the described person to see how it sets itself in motion and becomes the object of a more or less developed and complex sequence of events. To be sure, there is only a minimal indicator of this beginning action, but when the text opens a new paragraph saying "On commença la récitation des leçons" [We began reciting our lessons], we know that time has begun to move again. The brief freeze of time on the sudden image of a new pupil in class has given way to the usual procession of time. And as time moves on, the narrator's eyes remain on the new pupil, carrying the visuality of the initial description into the narrative.

It is quite remarkable that Flaubert chose to start his novel with such a self-conscious procedure of observation – a procedure that, as already stated, I take to be an important feature of the realist tradition. It is remarkable precisely because the procedure is not necessarily representative of the rest of *Madame Bovary*. Anyone familiar with Flaubert's novel will recall that the kind of perspective that we experience here at the outset of the book – seeing the events through the eyes of an onlooker, of a first-person-narrator who is a direct witness and participant of the narrated events – is discontinued already within the first chapter. Unlike the beginning, the rest of the novel is *not* told through the eyes of such a homodiegetic observer-narrator. Instead, the narrative voice shifts back and forth between that of a distanced, impersonal and omniscient narrator (*zero focalization*) and that of a more internally focalized style that explores the interiority of various characters through free indirect speech. It is thus only at the beginning that the narration is explicitly marked as based in an act of seeing. This disappearance of the homodiegetic narrator has, I argue, a lot to do with the difficulties in the initial procedure of observation.

Indeed, the first-person narrator and observer famously vanishes after this opening scene, reappearing only once more, still in the first chapter, to pronounce his inability to relate the rest of the story: "Il serait maintenant impossible à aucun de nous de se rien rappeler de lui." (Flaubert 1998, 19) [It would be impossible, now, for any of us to remember anything about him. (Flaubert 2004a, 10)] The fact that the homodiegetic narrator (the classmate and observer from the first scene) is unable to fulfill his role is, in some sense, sufficiently explained by the subsequent characterization of the protagonist, Charles Bovary – the very average man, who lacks any particularly striking features and does everything as he is supposed to: "C'était un garçon de tempérament modéré, qui jouait aux récréations, travaillait à l'étude, écoutant en classe, dormant bien au dortoir, mangeant bien au réfectoire." (Flaubert 1998, 19) [He was an ordinary sort of boy, who played during recess, worked during prep, paid attention in class, slept well in the dormitory, and ate well in the refectory. (Flaubert 2004a, 10)] Truly, such a man might be hard to remember, and his boring life may defy conventional narration. But this difficulty to remember and narrate Charles's life is also already thematic in the opening observation in which precisely the motion for which the narrator-observer waits remains strikingly limited: Charles just sits there, "n'osant même croiser les cuisses, ni s'appuyer sur le coude" [not daring even to cross his legs or lean on his elbow]. Charles Bovary, to whom we are introduced here, is indeed an awkward hero, disappointing immediately the onlooker's expectation of action. When the bell rings that ends the lesson, it takes a special admonition by the teacher just to have Charles fall in line with the others. The expected movement remains minimal, straining the patience of any observer.[3] Emphasizing Charles's feeble motion, Flaubert, in a way, already complicates the procedure of observation that I am interested in. The action that is supposed to follow the description sets in only very reluctantly. But the pattern of observation remains recognizable, and it seems significant that Flaubert chose to open his novel with such a pattern, allowing for a moment of description to bring a character into our view *and* setting this character in motion, introducing us thus to the novel's (second) protagonist (who will soon be replaced by the titular hero, Emma Bovary). Flaubert thus cites at the outset of his novel the realist technique of observation, but he also uses this technique of observation to show the particular challenge faced in his work: the challenge to tell the story of a man who has no story. Charles Bovary is a man whose appearance can be described, but whose boring life resists narration.

Observation and literary realism

The opening scene of *Madame Bovary* stands here as an example of a literary technique that I call, borrowing from the history of science, an 'observation.' Literary observations consist in the combination of description and narration; they stage the transition from the static description of a newly appearing person (or object) to the narration that captures how this person (or object) acts as time passes. Such observations, I argue, are a crucial (but by no means exclusive)

---

3 My argument here slightly varies from Jonathan Culler's reading of this scene. Culler, too, emphasizes that the observer's inability to serve as narrator is captured already in the opening sequence of the novel. However, instead of focusing on Charles's lack of motion, Culler highlights the fact that the text points us from the outset to the impossibility of the description we are reading: what we read cannot actually be the result of what the homodiegetic narrator himself sees. Indeed, while the detailed description appears to derive from an observer within the room, this same observer tells us that Charles stood behind a door so that one could hardly see him (Culler 2007, 692–693).

trait of literary realism. However, they also tend to complicate what we often take literary realism to be. Essentially, these observations complicate a definition of realism that is based on a strict opposition between description and narration and that emphasizes realism's preference for description.

Under description I understand here the representation of static images or *tableaux* of landscapes, interiors, clothes, and physiognomies. Narration, in contrast, is the representation of actions and sequences of events. This definition follows Gérard Genette's canonic definition of description and narration in his essay "Frontières du Récit" [Frontiers of Narrative]. As Genette stresses, narration is the representation of actions unfolding over time; description, by contrast, deals with the representation of persons and objects in space and it represents states rather than sequences:

> [L]a narration s'attache à des actions ou des événements considérés comme purs procès, et par là même elle met l'accent sur l'aspect temporel et dramatique du récit ; la description au contraire, parce qu'elle s'attarde sur des objets et des êtres considérés dans leurs si-multanéité, et qu'elle envisage les procès eux-mêmes comme des spectacles, semble suspendre le cours du temps et contribue à étaler le récit dans l'espace. (Hamon 1991, 262)

> [N]arration is concerned with actions or events considered as pure processes, and by that very fact it stresses the temporal, dramatic aspect of the narrative; description, on the other hand, because it lingers on objects and beings considered in simultaneity, and because it considers the processes themselves as spectacles, seems to suspend the course of time and to contribute to spreading the narrative in space. (Genette 1982, 136)[4]

It is, of course, a relatively weak and broad notion of narration that I use here when I define narration so loosely as the representation of a temporally extended sequence. Although my emphasis on temporality certainly captures one important aspect of what theorists generally take narration to be, it ignores another aspect that figures prominently in definitions of narration. As the German narratologist Wolf Schmid summarizes, narrative (the object of narration) is generally defined not only by temporality, but also by a 'change of state' – Schmid himself prefers here to speak of an event (*Ereignis*) at the core of narration (Schmid 2003). To be clear, I will not pay much attention to the criterion of a change of state. The object of narration, in this study, will be simply a temporally extended chain of actions. In defense of this 'weak' notion of narration, it can be said that a temporal development always implies some change of state, however minimal this change may be. Time leaves nothing untouched.

---

4 This distinction between description and narration is, to some extent, echoed in Wolf Schmid's book *Narratology: An Introduction* (Schmid 2010, 5–6).

The distinction between description and narration has crucially informed debates on the poetics of literary realism. What I am especially interested in is the longstanding argument that links description to realism because of description's potential to make us see the world, to put the world before our eyes. The idea that description can make us see the things described can already be found in Quintilian's *Institutio Oratoria* [The Orator's Education] (around 95 CE), and Quintilian influences a long tradition of this thought from early modernity to the present. Making us see an object through description is in this tradition, and especially from the eighteenth century onwards, increasingly understood as a realist technique: evoking the world 'as it is.' Description, in this tradition, is, more precisely, not only that technique that makes us see the world; it is – particularly in the nineteenth century – also staged as a *seeing* of the world. Descriptions often appear in texts when a character looks out of the window, or down from the top of a hill, or as he/she walks through the city. Description constructs a visible image of the world as it was (or is) actually seen. Moments of descriptions in novels, novellas, and short stories are moments of vision – for the characters and narrators as well as for the readers.

According to this understanding, realism becomes very much a product of description alone. It is in the descriptions that the true and detailed fabric of the world becomes palpable (or visible, I should say). Narration, by contrast, simply places a plot within this visual reality. But this is a position that is hard to accept for all those scholars who identify literature's potential to represent reality with its production of narrative, with literature's ability to narrate the actions of men and women. Description, according to these thinkers, can offer us only disconnected still lifes. The most well-known representative of this position today is Georg Lukács, who, in 1936, published an essay with the simple title "Erzählen oder Beschreiben?" [Narrate or Describe?]. But the critique of description and the foregrounding of narration are much older, going back at least to the eighteenth century – to Lessing in Germany and Boileau in France.

What all of this amounts to is that there exists in the field of literary theory a peculiar tension between those who link realism to description because of description's visuality and those who link realism to narration because of narration's ability to represent coherent chains of human action. To be sure, in the rich scholarly literature on this topic one also finds a range of still other positions – and I will consider at least some of these in this book (especially in the following chapter, which offers a more detailed survey of the debates on description and narration). For the moment, however, let us just acknowledge a general tension between the theorists of a 'descriptive realism' and those of a 'narrative realism.'

My own intervention consists in highlighting and dissecting a specific technique of literary realism – I call this technique 'observation' – in which the visuality of description is carried over into the narrative. Observations rely, first of all, on the possibility to interrupt a narrative and describe a person (or object) in some detail, but they ultimately aim to extend the visuality of the inserted description into the narrative. Observations thus complicate a notion of literary realism whose insistence on visuality is linked to description alone. Observations construct a world that is wholly visible and 'real,' not just in its still-lifes of descriptions, but also in its narrative sequences. Observation thus offers us a paradigmatic solution to the problematic relation between description, narration, and literary realism.

I do not proceed in this book through the analysis of countless short and isolated observations, like the one we saw at the beginning of *Madame Bovary*, interesting as this might be. Instead, I explore a rather heterogeneous group of novels, novellas, and short stories from the early Enlightenment to the eve of Modernism in which the procedure of observation is, in one way or another, turned into a central concern of the entire text. These texts are not necessarily themselves canonical works of what we conventionally understand to be the period of literary realism. Some of the texts that I have chosen are too old for this period (such as Lesage's *Le Diable boiteux* [The Staggering Devil], 1707), while others are too new (such as Doyle's *Sherlock Holmes*, 1887–1927). Again others, like Goethe's *Sturm und Drang* novel *Die Leiden des jungen Werthers* [The Sorrows of Young Werther] (1774), just do not fit easily at all into the development of literary realism. However, all of these texts, as my readings aim to show, contribute some important aspect to our understanding of the technique of observation. All of these texts interestingly complicate what appears to be such a simple transition from the description of a new object or person that has (more or less suddenly) come into view to the sustained watching that captures how this object or person is set in motion. These texts help us explore how a narrative sequence of events gives way to description, and how the visuality of description is then carried over into narrative. Moreover, these texts show how the narrative procedure of observation is affected by a range of discourses and cultural techniques – from the natural sciences through the mutual surveillance in the emerging big cities to art history. In other words, this book contributes to our understanding of the formal and cultural underpinnings of literary realism by way of a detour, through readings of texts that themselves do not necessarily belong to the realist tradition.

Speaking of all the different discourses that intersect with the narratological concerns of description and narration, I should now also explain why I call the procedure of seeing that combines description with narration 'observation.' I use the term observation here because the literary technique that I study corresponds in interesting ways to the contemporaneous scientific procedures of observation, which also deal in many cases with the combination of individual images with dynamic sequences. I present in this book a number of such scientific observational procedures, and although these procedures all work differently from one another in important regards, they are all crucially defined by the struggle with the combination of image and sequence.

However, before I develop my understanding of scientific observation in more detail in the following section, I should repeat that the readings in the chapters of this book will lead me to consider a wide range of discourses, not all of them scientific. And it is important to me not to reduce the literary problem of observation to a mere after-effect of the history of science (this would be a claim that is neither tenable nor interesting). But I do argue that an awareness of the practice of scientific observation, which is of great importance to the two centuries that I study in this book, allows us to appreciate the difficult transition from the seeing of images to the seeing of sequences as a central concern also of literary narratives.

Scientific observation

Let us here recall the great rise of observation as a method of scientific inquiry from the late seventeenth century onwards, which Lorraine Daston and others have impressively reconstructed in recent years. With the early eighteenth century – in the wake of the scientific revolution and its new emphasis on empiricism – observation became the dominant method for the production and testing of knowledge in a wide range of disciplines, many of them newly emerging. While observation was previously restricted to medicine and astronomy, or even to such a "lowly art" (Daston 2008, 102) as the everyday-meteorology on the farmlands, it was now defining the work in biology, geology, psychology, economics, and many other fields. In her 2008 essay "On Scientific Observation," Lorraine Daston highlights this great ascent of observation:

> Far from being a lowly art, plied by unlettered artisans and peasants, as it had been regarded earlier, or an inferior substitute for experiment, as it was later viewed, observation had by the early eighteenth century become an essential and ubiquitous scientific practice, an art in the service of science. (Daston 2008, 102)

Interestingly, Daston emphasizes in this account of the rise of observation also another scientific practice with which observation historically competed and which is important for us to understand: experiment. Starting in the late seventeenth century, observation is increasingly seen as superior to experimentation; but this view changes again in the course of the nineteenth century, which sees the eventual triumph of experiment. I will return to the opposition between observation and experiment more extensively later on in this book (notably in my reading of *Les Nuits de Paris* [The Nights of Paris] in the third chapter). But the main difference between experimentation and observation is that in contrast to the experimenter, who determines the framework of his investigations and seeks to prove or disprove a hypothesis, the observer remains essentially open "to possibilities for new knowledge in the most unexpected places" (Daston, Lunbeck 2011, 8). This peculiar openness of the observer, I will argue in more detail below, is at the origin of his need to be simultaneously distractible and focused, to see sudden new images and to focus on an evolving sequence. The experimenter, by contrast, can focus from the very beginning on a given sequence.

The breeder of domestic animals can serve us here as a first example of the figure of the scientific observer. Extensively analyzed at the outset of one of the most important science books of the nineteenth century – Charles Darwin's *On the Origin of Species* (1859) – the breeder has a claim to special importance in the history of knowledge. What makes the figure of the breeder, moreover, particularly relevant to the present study on literary observation is the fact that the professionalization of breeding happened in roughly the same period in which the modern novel emerged and in which observation reigned supreme in the sciences: from the mid-eighteenth to the mid-nineteenth centuries.[5]

To appreciate what characterizes the breeder in his capacity as an observer, consider the following passage from Darwin's *Origin*. Regarding the fine gaze necessary for the work of the breeder, Darwin writes admiringly:

> If selection consisted merely in separating some very distinct variety, and breeding from it, the principle would be so obvious as hardly to be worth notice; but its importance consists in the great effect produced by the accumulation in one direction, during successive generations, of differences absolutely inappreciable by an uneducated eye—differences which I for one have vainly attempted to appreciate. Not one man in a thousand has accuracy of eye and judgment sufficient to become an eminent breeder. (Darwin 2008, 27)

To be sure, one thing that Darwin strongly highlights in his explanation of the work of the breeder is the breeder's ability to perceive utterly small differences.

---

5 When Darwin published *On the Origin of Species* in 1859, the "methodical practice" of breeding had been in place "for scarcely more than three-quarters of a century" (Darwin 2008, 28). To be sure, less methodically breeding had been around for much longer – but the same could be said for the novel.

This emphasis on detail in nineteenth-century science is familiar enough. Scholars have amply discussed how the preoccupation with detail defined Darwin's century and left its imprint also on the development of the realist novel from Balzac and Dickens to Stifter and Zola. What we see in this period is an emerging awareness of the importance to stop and describe all those small things hitherto ignored. But what Darwin really emphasizes here is not only the description of detail, but also the temporally extended gaze of the breeder that captures the "great effect produced by the accumulation in one direction, during successive generations." The breeder does not merely describe a state: he/she apprehends a minute deviation in one individual and then follows its development over a long time. The breeder does not merely describe a state of being; he/she carries the description over into the registration of a prolonged development. The breeder is thus engaged in two distinct, albeit connected, operations. He/she is ready to be struck by any minute deviation in a new generation, and he/she is able to focus on the sustained preservation and development of a single detail over a long time. The breeder sees momentary pictures *and* extended sequences.

Implied in this temporal duality of observation, which captures both state and development, is another duality, one that concerns attention. For while the readiness to be struck by any minor deviation in a newborn individual requires a state of perpetual distractibility, the prolonged recording of the preservation and development of a specific trait over many generations demands a sustained narrow focus. Observation, in my definition, thus does not fit easily into any narrative of modernity as an age of either increasing attention or of distraction.[6] It is certainly true that observation also requires a good deal of attention and is thus dependent on a new emphasis on the prolonged attention to everyday phenomena that was developed over the course of the seventeenth and eighteenth centuries – a process elegantly summarized by Lorraine Daston in her brief essay *Eine kurze Geschichte der wissenschaftlichen Aufmerksamkeit* [A Brief History of Scientific Attention] (2000) and elsewhere (Daston 2004). Daston recalls, for instance, the case of the eighteenth-century scholar Charles Bonnet (1720 – 1793), who "observed a solitary aphid every day for about a month from 5:30 a.m. to 11:00 p.m., duly recording all data in a table" (Daston 2004, 437; see also Daston 2000, 34 – 35). Through examples like that of Bonnet, Daston reconstructs the "disciplines of attention," which served to "single out objects as worthy of sustained investigation, isolate them from the continuous flow of ex-

---

6 The abundant debates on the problems (or possible benefits) of distraction in modernity have been the subject of a number of significant studies in recent years; see for, instance, Crary 1999, Daston 2000, Daston 2004, North 2011, Phillips 2016.

perience, [and] subject them to an exacting mental, often literal, dissection" (Daston 2004, 436). But what Daston here summarizes under the rubric of attention seems, in fact, to be a more complex procedure. Instead of just requiring attention, observation confronts the subject with the paradoxical double imperative both to focus and to be distractible: to be open to new knowledge in unexpected places and to attentively watch an object that has come into focus. To summarize then, the concept of observation that I introduce in this book is a complex compound of two different kinds of seeing: a seeing that is both focused and distractible, directed both at momentary pictures and at extended filmic sequences.

Darwin's breeder is open to being struck by newly emerging deviations in unexpected places, and he/she is able to focus on the preservation and development of a specific trait over many generations. Similar – albeit certainly not identical – dualities of seeing can be recognized in a wide range of other observational disciplines of the eighteenth and nineteenth centuries. Be it in the observation of the finite velocity of light, as discussed by Rømer and Bradley in the seventeenth and eighteenth centuries (see Bobis and Lequeux 2008, Cohen 1940, Sakellariadis 1982), which was too fast to be directly witnessed, or in the observation of the evolution of species (see Förster 2001, esp. 93) or mountain structures (see Rudwick 2010), which is too slow to be easily seen, observation often struggles with the complex negotiation of state and sequence. In all of these cases, the observer is open to the transition from the seeing of states to the seeing of sequences or even to discover a sequence where there is seemingly merely a stable state.

Let us read here a passage from *Principles of Geology* (1830–1833), in which Charles Lyell (1797–1875) praises the "result of observations" conducted in the field of geology over the preceding thirty years. Lyell stresses that the achievement of geological observations was, similar to that of the more established astronomical observations, to have revealed movement and development where there was apparently none:

> The senses had for ages declared the earth to be at rest, until the astronomer taught that it was carried through space with inconceivable rapidity. In like manner was the surface of this planet regarded as having remained unaltered since its creation, until the geologist proved that it had been the theatre of reiterated change, and was still the subject of slow but never ending fluctuations. (Lyell 1997, 24)

Lyell's statement is an index of the great awareness of eighteenth- and nineteenth-century scientists that their observations revolved around the seeing of dynamic sequences and stable states. Negotiating between state and sequence defined the scholarly research even in subjects in which the observed movement was at a tolerable pace, as, for instance, in the movement of clouds. Here, too, observation necessarily has to start from the initial definition and description of static cloud forms in order to capture continuous movements (on the difficult process of agreeing on such cloud forms, see Daston 2008, 103–105).

No two of these scientific procedures, it bears repeating, are identical. The breeder's transition from the identification of a deviation in a newborn individual to the sustained recording of the development of this deviation over many generation is very different from the geologist's developing realization of the ways in which the history of the earth is inscribed in the currently visible landscape. And both of these endeavors are crucially distinct from the meteorologist's attempt to define stable cloud forms that allow him/her to describe the ongoing movements in the sky. But what all of these kinds of observation have in common is that they struggle, in some form or other, with the problem of the transition from the description of a state of being to the recording or reconstruction of an extended sequence. And it is on this general, but meaningful level that fictional texts from the early eighteenth century to the late nineteenth century reflect the development of observational procedures. The duality in the temporality of scientific observation – which includes the seeing of both image and sequence, of distractibility and focus – is reflected in literature in the negotiation between static description and dynamic narration.

My intention in this book is not to derive from the history of science a direct model for the analysis of eighteenth- and nineteenth-century novels. Neither do I aim to reveal – inversely– that inscribed in the scientific development of observational procedures are the poetics of description and narration. Instead, my interest is to suggest a certain array of family resemblances between problems of literary representation and scientific research in the eighteenth and nineteenth centuries and to sharpen our eyes for observational procedures in literature through a comparative glance at the history of science. The argument of a co-evolution of observation in science and literature is thus different from Shapin and Shaffer's influential view on "literary technologies" in *Leviathan and the Air-Pump* (1985). While for Shapin and Schaffer the literary method is an immediate part of the scientific process, i.e. the description of experimentation (Shapin, Schaffer 1985, 22–79), I develop my argument about literary observation here based on the assumption that scientific and literary devices develop in a parallel way and roughly during the same period and yet in an autonomous manner.

Observation in science and literature

The comparison between observations in literature and the sciences helps us discover interesting affinities and parallels between these two fields of knowledge, even for a century – such as the nineteenth – in which science and literature started to diverge significantly. But the comparison between observations in science and literature does not only illuminate curious parallels, but also sharpens our eyes for what is specific about literary discourse. To understand how literary observation remains distinct from scientific observation, I have to turn here to another definition of narration – one that has very little to do with the distinction between narration and description on which I generally rely in this study.

As Wolf Schmid lays out in his essay "Narrativity and Eventfulness" (2003), there are largely two competing definitions of the term narration. These two definitions lead to almost entirely different concepts of narration, and yet it makes sense to take both of these definitions into account in the analysis of a literary text, as is indeed often done. According to the first definition, narration is contrasted with description. Description represents static images; narration, in contrast, represents temporally extended sequences (and usually a change of state). This is the definition that I use to explain observations in general, as procedures that combine description with narration. But there is also another definition of narration, and according to this second definition, narration is characterized by the presence of a mediating voice (namely, that of the narrator). In this definition, narration is contrasted not with description, but, instead, with the unmediated representation of action as, for instance, in drama (where characters speak directly, without the mediating presence of a narrator).

I concur with Wolf Schmid that it is useful to combine these two definitions of narration, but also to keep them terminologically apart. Schmid thus distinguishes, on a first level, between narrative and descriptive texts. On this level, a drama, too, would be a narrative text, as it represents actions that unfold over time (and which usually contain a change of state). On a second level, however, Schmid distinguishes between narrative texts that are "showing" (like drama) and narrative texts that are "(story-)telling." Schmid defines these storytelling narrative texts as narrative texts in a narrow sense. (To be clear, all the texts discussed in this book are narrative texts in this narrow sense. I will be dealing with novels and other, shorter forms of fiction.) This inclusion of storytelling into the definition of narration has, as I will argue in more detail in my reading of *Les Nuits de Paris* (chapter 3), major implications for the thinking of the relation between scientific observation and literary observation. The presence of the mediating voice of the narrator always suggests a certain organization of the related materials. This is especially clear if the text is presented in the past tense: the events are 'over' and what is presented to us has been deemed worth telling by a narrator whom we trust to tell us only what we need to know to appreciate the story. But even if the story is told in the present tense, the presence of the narrating voice promises some degree of structure and organization. As Andreas Kablitz puts it in his essay "Realism as a Poetics of Observation": "Narration is retrospective in nature: it organizes time by moving backwards from the end of what has happened" (Kablitz 2003, 133).

This idea of an organizing structure of storytelling, however, heavily complicates the notion of 'openness' on which I rely in my definition of observation. Literary observations, I stated, are meant to perform for the reader a process of perception that moves from the openness to be struck by a sudden image to the focused watching of this image as it develops over time. But this idea of an openness to be struck by the sudden appearance of an image that is curious, unexpected, and new, and that might be worth watching over time is at odds with the reading of a literary text that is orchestrated by a mediating narrative voice. The narrator serves as the guarantor of the fact that everything we are told is worth noticing and that everything will fit together in an organic whole.

Such an organizing figure of a narrator is – at least at first sight – absent in scientific observation. In contrast to literary storytelling, which organizes the events through its retrospective gaze, scientific observation is an essentially open and future-oriented process. Andreas Kablitz puts it thus: "Observation [...] is forward-looking; thus, it does not function in pursuit of an objective but, precisely because it does not know any such objective, finds its way as it goes" (Kablitz 2003, 133).

A novel can, of course, show characters who are open to new discoveries and who are willing to follow up on these new discoveries (and I will discuss such characters repeatedly in the following chapters); but that is different, one might say, from the performance of a process of observation for the reader. Unlike the characters in the text, the reader experiences the text as originating in an act of storytelling that guarantees a certain kind of closure in which everything is meaningful, and in which the shifting between distractibility and attention is not required, at least not to the same degree.

It might appear that I have now already arrived at a fairly clear notion of the similarities and differences between literary and scientific observation: while both literary and scientific observation deal with the connection of a seeing of images and of extended sequences, only scientific observation seems to be actually faced with the necessity to shift between two types of attention: the willingness to be distracted by a new image and the ability to focus on this image over time. Literary texts can portray characters who face the need to mediate between distractibility and prolonged attention in this manner; but literary narratives cannot easily perform this duality for the reader.

However, rather than adhering to this statement of a relatively clear distinction between literary and scientific observation, I want to suggest that literary observations aim precisely to overcome the distinction that I just outlined. In taking up the duality of two forms of seeing (the seeing of suddenly appearing images and the recording of a following sequence), literary texts in some way try to perform also a kind of openness to new images that is essentially alien to the closure implied by the act of storytelling. When we read the opening of *Madame Bovary*, for instance, we could, of course, already guess that whatever the novelist will place at the beginning of this novel will matter, and we can expect to be introduced here to an important character for the novel about whom we will hear much more as we continue reading. And yet Flaubert – particularly by mediating the appearance of Charles Bovary through the eyes of the schoolchild, who looks up to see a new boy appear in class – somehow induces us to tend to this scene not with the expectations of a reader of fiction (aware of storytelling's structure), but as an *observer*, curious about this newly appearing sight of Charles Bovary, and willing to keep watching. Simple as the procedure of literary observation appears, it thus actually works at the limit of its own medium of storytelling, creating an effect of openness to unexpected images worthy of further study.

I am very much sympathetic also to the objection that my opposition between the epistemology of scientific observation and literary storytelling remains still too stark – my qualification in the last paragraph notwithstanding. In science, too, the openness to new phenomena is certainly much more difficult to accomplish than my remarks suggest. It hardly needs reference to the insights of cognitive sciences, classic phenomenology, or Luhmannian systems theory to understand that the observer is able to see only those phenomena that fit a pre-given code. Nothing radically new can ever be perceived. The ideal of the openness of the observer to new phenomena is at odds with the actual restrictions on human perception and knowledge. Conversely, it is easy enough to contest also my broad statements about the closure guaranteed by the narrator. In some way, much of the development of the novel in the nineteenth century works precisely at the deconstruction of the image of the authoritative voice governing over the communication of the story. Instead, we are increasingly shown the world through the limited perspective of individual characters and are given ample time to contemplate not just the narrative development along a clear arch, but also the long fillers that showcase the prose of the world (a development discussed in some detail in Franco Moretti's 1987 study *The Way of the World*). The guarantee of a closure of storytelling and of the meaningfulness of everything we are told is, in other words, increasingly called into question. And yet, despite these very legitimate objections, it seems that the category of openness remains useful to distinguish between the basic epistemological situations in science and in literary storytelling. It would be a strange misrepresentation to suggest that the scientific study of nature proceeds under the same conditions of closure as literary storytelling. Literary observation, in a framework of storytelling, is less open than scientific observation; this distinguishes it from scientific observation. However, this distinction is neither absolute nor static; in fact, literary observations are a means to overcome precisely the expectations set by the closure of storytelling.

Observation and pragmatographia

At this point, I have to go back a few steps to clarify another element of my concept of literary observation. I said that literary observations combine description with narration, that they emphasize the possibility to focus on a given image *and* to carry its visuality over into the narrative. I defined description as the representation of static, temporally not evolving images, and narration as the representation of how people (or objects) develop over time. Description, I argued further, visualizes to the extent that it focuses on the fabric of the world, giving us a picture of what things look like. However, I neglected to fully acknowledge that my definition of description as static is by no means universally accepted. In fact, some theories of description explicitly include temporally extended phenomena.

Indeed, from a certain perspective, the definition of literary observation as a form of description that exceeds individual images to include temporally extended phenomena sounds like a very familiar literary tool. Already in early modern rhetorical treatises, we find among the many forms of description – from *topographia* (the description of a place) and *prosopographia* (the description of someone's face or character) to *dendrographia* (the description of trees) and *hydrographia* (the description of water) – also a kind of description that seems to correspond to my notion of literary observation, namely *pragmatographia*, the description of actions. George Puttenham, for instance, defines in his influential book *The Arte of English Poesie* from 1589 pragmatographia as the "the Counterfait of action" (Puttenham 1869, 246):

> But if such description be made to represent the handling of any busines with the circumstances belonging thereunto as the manner of a battell, a feast, a marriage, a buriall or any other matter that lieth in feat and activitie: we call it then counterfait action (*Pragmatographia*). (Puttenham 1869, 246–247)

While pragmatographia resembles observation in its ability to capture temporally extended phenomena, the procedure of observation that I study in this book still differs in important ways from this traditional rhetorical device. Observation is most importantly different from pragmatographia in that it explicitly engages the relation between the description of individual images and the narration of sequences. Observation is that procedure that reaches from the description of a static image to the focused scrutiny of a sequence of events occurring over time. In contrast to pragmatographia, observation originates in the complex combination of moment *and* sequence – description *and* narration. The combination of these two stages in the observational process – the seeing of an image and the seeing of a sequence – is a necessary trait of observation because literary observation, like the scientific practice of observation, performs a form of perception that remains open to "new knowledge in the most unexpected places" (Daston, Lunbeck 2011, 8). Because the unprejudiced observer is, at the outset of his/her inquiry, open to attend to whatever strikes his/her attention, he/she is initially limited to the registration of an image. Seeing a sequence would require that he/she already focuses on one image. Only once he/she is struck by an image can he/she focus on this image and begin to observe a sequence. Being struck by an image and focusing on a sequence are the necessary stages in which observation proceeds. Observation is the complex compound of the readiness to be struck and distracted by anything new, and the temporally extended focus on a specific element; a compound of the seeing of an image and of an extended sequence.

One consequence of this transformation from image to sequence is a form of 'zooming-in,' a narrowing of the gaze on one specific object, person, or situation. It may be possible to describe a stable state of being in all its circumstances, but once we follow a sequence of events, we have to decide which object we want to keep in view (until we are distracted again by something else). Tellingly, early modern rhetorical handbooks that deal with the device of pragmatographia do not consider this zooming-in at all. Quite to the contrary, the leading rhetorical treatises of the early modern period emphasize that the writer who uses pragmatographia should offer a panoramic view of all that happens. This stress on a circumstantial account is apparent in Puttenham's definition of pragmatographia, and we find it in even more pronounced form in Henry Peacham's much-consulted *The Garden of Eloquence* (1577). Consider here the long list of simultaneous events that Peacham instructs the writer to describe when giving an account of a city under assault (the example itself is inherited from Quintilian[7]):

> Pragmatographia, a discription of thinges, wherby we do as plainly describe any thing by gathering togeather all the circumstaunces belonging unto it, as if it were moste lively paynted out in colloures, and set forth to be seene, as if one should say, the citty was overcome by assault, he compryseth al in a summe. Whatsoever such fortune suffered, as sayeth *Fabius*, but if thou wilt open and set abroade those thinges whiche were included within one word, there shall appeare many fyres and scattered flames upon houses and temples, the noyes of houses falling down, one sounde of divers thinges and cryes, some flye with great daunger, others hang on their friendes, to bid them farewell for ever, the scriking of Infantes, women weepinge most bitterly, old men kepte by most unhappy destiny to see that day, the spoyling of temporall and hallowed thinges, the running out of them, that caryed awaye spoyles, and of them that intreated for their owne goods, every man ledde chayned before his spoyler, the mother wrastling and stryving to hold her sucking babe, and whersoever were great riches, there was great fighting among the spoylers. (Peacham 1971, no page number, emphasis in the original)

Peacham's account of how to describe the assault on a city emphasizes the many particular aspects that can be part of such an assault. It is, as noted above, an insight known since antiquity that such mentioning of particulars serves to visualize a scene (see Innocenti 1994).[8] It becomes thus, with some important qualifications (more on these below), possible to visualize temporally extended phenomena.[9] And yet there is a striking difference between pragmatographia and observation. Most importantly, pragmatographia manages to visualize temporally extended phenomena by offering individual flashes or snapshots of a variety of particular images. But we do not see any of these images themselves actually develop. Even if the object of pragmatographia (in the example above, a city under assault) is temporally extended, we do not see any of its individual ele-

---

7 See Quintilian 1976, 248. Peacham refers to Marcus Fabius Quintilianus in the quotation below by the name of Fabius.

8 Other elements of the classic poetics of vivid description are clearly drawn on in this scene as well, notably the usage of contrast. Consider here, for instance, the contrasting images of "the scriking of Infantes" and the "old men kepte by most unhappy destiny to see that day." However, in some way, such contrasting is really just part of the strategy to emphasize particular images, because the images will appear even more 'particular' through the contrast to other, opposed images.

9 Because of this emphasis on visualization, pragmatographia is distinct from simple narration – from what Aristotle in his *Poetics* called *synthesis pragmaton* [combination of actions]. Pragmatographia is not simply the combination of actions; instead, it is a specific technique to visualize temporally extended phenomena by highlighting a variety of distinct particular images.

ments really change over time. Observation, by contrast, is concerned precisely with this transition from a suddenly appearing image to its extended watching. Where pragmatographia offers a spectacle of striking images, observation focuses on one of these images to see how it develops. Observation requires that we be open to being struck by an image and have the attention to follow this image over time. Pragmatographia, by contrast, captures at best only this initial stage of distractibility: left and right we are distracted by individual images of a city under assault, but nothing is truly observed. The fact that pragmatographia captures temporally extended phenomena is thus, in some sense, almost meaningless, because pragmatographia does not focus on any individual aspect long enough to capture a real sequence.

Looking for an example of the difference between pragmatographia and observation, Flaubert proves, once more, very instructive. Indeed, the opening of his novel *L'Éducation sentimentale* [Sentimental Education] (1869) shows this difference quite clearly. Famously, this novel opens with the departure of a ship from the port in Paris. The description of this departure – a temporally extended process – is a clear case of pragmatographia:

> Des gens arrivaient hors d'haleine ; des barriques, des câbles, des corbeilles de linge gênaient la circulation ; les matelots ne répondaient à personne ; on se heurtait ; les colis montaient entre les deux tambours, et le tapage s'absorbait dans le bruissement de la vapeur, qui, s'échappant par des plaques de tôle, enveloppait tout d'une nuée blanchâtre, tandis que la cloche, à l'avant, tintait sans discontinuer. (Flaubert 1983, 3)

> People arrived, out of breath; barrels, ropes and baskets of washing lay in everybody's way; the sailors ignored the passengers; the baggage was piled up between two paddle wheels; and the general din merged into the hissing steam, escaping through some iron plates, wrapping the whole scene in a whitish mist, while the bell in the bows was clanging incessantly. (Flaubert 2004b, 5)

The text highlights various images belonging to the scene of a boat leaving the port: people are rushing, baskets and cables are in the way, the steam is hissing, the bell is ringing. But none of these images is followed up in any way; all these images just flash up briefly to contribute to the visualization of the process that is under way: a ship preparing for departure.

In a next step, however, the text radically changes pace. Suddenly a single image comes into view and receives more detailed attention. It is the image of the protagonist, Frédéric Moreau, standing near the tiller, literally motionless, as if posing for this 'picture':

> Un jeune homme de dix-huit ans, à longs cheveux et qui tenait un album sous son bras, restait auprès du gouvernail, immobile. A travers le brouillard, il contemplait des clochers,

des édifices dont il ne savait pas les noms ; puis il embrassa, dans un dernier coup d'oeil, l'île Saint–Louis, la Cité, Notre-Dame ; et bientôt, Paris disparaissant, il poussa un grand soupir. (Flaubert 1983, 3)

A long-haired young man of eighteen, holding a sketchbook under his arm, stood motion-less beside the tiller. He gazed through the mist at spires and buildings whose names he did not know, took a last look at the Île Saint-Louis, the Cité, and Notre-Dame; and soon, as Paris was lost to view, he gave a deep sigh. (Flaubert 2004b, 5)

From that point on, the text follows this young man as he walks along the ship, and, more generally, as he acts as the protagonist in this novel. Out of the initial mass of snapshots that constitute the pragmatographia thus crystallizes the one image of the protagonist who is now followed over time. Flaubert's novel thus performs the transition between the merely distracted spectatorship of pragmatographia to the process of observation.[10]

Competing concepts of observation

Before closing this introductory chapter, I should make clear how my definition of observation as the combination of description with narration relates to other recent attempts to theorize observation, and I should also say a few words about the content of the remaining chapters.

For many literary critics of the twentieth and twenty-first centuries, observation remains a rather loose concept, predominantly associated with description. Both Georg Lukács and Peter Brooks, for instance, use the term observation simply as a synonym for description or as a vague reference to the activity that precedes the actual linguistic act of description.[11] Neither Lukács nor Brooks discusses the origin of the practice of observation in the history of science; nor do they analyze the mechanisms by which observation actually functions in individual literary texts.

But even for literary scholars who comment more explicitly on the connections between scientific practice and literary form, observation and description sometimes remain closely allied. In his study of the Victorian Novel, Louis

---

10 In a kind of *mise en abyme*, the novel subsequently repeats this transition from pragmatographia to observation in the sights taken in by the protagonist: he, too, soon transitions from the distracted seeing of a multitude of different buildings along the river to the focus on one single man who is on the boat with him.

11 Brooks and Lukács speak of "Beobachtung und [...] Beschreibung" (Lukács 1955, 136) [observation and representation (Lukács 1970, 147)] and "observation and representation" (Brooks 2005, 71) See the discussion of Brooks and Lukács in the following chapter.

James, for instance, draws a direct connection between the rise of "scientific observation" since the eighteenth century and the literary "description" of things, "as they really are" (James 2006, 29).[12]

Over the course of recent years, however, some critics have considered the concept and practice of observation more closely – which reflects a generally growing interest in the epistemological affinities between the sciences and arts. But the definition of observation in these more recent studies remains still different from the one that I present here. Most importantly, perhaps, the procedure of observation that I study in this book departs from the procedures that Jonathan Crary analyzed in his influential 1990 book *Techniques of the Observer: On Vision and Modernity in the Nineteenth Century.* While Crary is also interested in hitherto ignored parallels between artistic and scientific notions of seeing in the nineteenth century, his primary concern is with a process in which vision was increasingly understood to depend on the bodily conditions of the viewer. What Crary wants to work out in his book is "the way in which concepts of subjective vision, of the productivity of the observer, pervaded not only areas of art and literature but were present in philosophical, scientific, and technological discourses" (Crary 1990, 9). Crary's project to showcase the heightened awareness of subjectivity in a wide range of fields of knowledge in the nineteenth century has been very successful and has inspired several subsequent studies on vision and subjectivity.[13] Subjectivity should be understood here not in the everyday-sense that what we see is 'merely subjective' and thus in some sense up to each individual viewer. Instead, Crary is interested in the ways in which scientists and artists became increasingly aware of the ways in which vision depended on the body of the subject – on the specific physiological makeup of the eye. The study of afterimages in Goethe's *Zur Farbenlehre* [On the Theory of Colors] (1810) serves Crary as one important example of the increasing awareness of the physiological preconditions of vision. The research into these images, which appear after one has closed one's eyes, implies an awareness of the fact that images are not only a reproduction of an exterior reality, but also

---

12 James references here Raymond Williams (see Williams 1976, 217). See also Deborah Shapple Spillman's comment on the "highly descriptive" (Shapple Spillman 2012, 136) late nineteenth-century novel *The Farm in Karoo* by Mary Ann Carey-Hobson: "Encountering the diversity of the natural world and ways of naming it, characters and readers alike learn to observe, describe, and classify the world as specimen" (Shapple Spillman 2012, 136).

13 The prominence of such work on the subjectivity of vision in recent Goethe studies, for instance, is mentioned by Piper (Piper 2010, 30). A recent publication in this field, also mentioned by Piper, is the volume *The Enlightened Eye: Goethe and Visual Culture* (2007), edited by Evelyn K. Moore and Patricia Anne Simpson.

a product of the subject itself: "The privileging of the afterimage," Crary writes, "allowed one to conceive of sensory perception as cut from any necessary link with an external referent" (Crary 1990, 98). A similar interest in the production of images in the subject motivated, according to Crary, the scientific study of binocular disparity – the phenomenon by which we see a slightly different image with each eye. For the understanding of binocular disparity implies that no image that we believe to see 'out there' when we have both eyes open directly corresponds to any immediate perception of an exterior reality. What we see is, instead, a "a conjuration, an effect of the observer's experience of the differential between two other images" (Crary 1990, 122). As scholars began to appreciate in the first half of the nineteenth century, binocular disparity is of central importance to the way we see the world. Only thanks to the combination of the two different images do we have any depth perception and can we easily relate objects to each other in a three-dimensional space.

In some sense, the present study borrows little from Crary's emphasis on the evolving insights into the importance of the body in the production of vision. In contrast to Crary, I take an enduring belief in the disclosure of reality through vision, with all due qualifications (and Crary already points us to some of the most important), to be an important underlying assumption of the eighteenth and nineteenth centuries (I will have a bit more to say about this in the following chapter as well as in the conclusion). And yet my line of inquiry is maybe not all that far from Crary in that the readings in this book will show us repeatedly that the disclosure of reality through observation is nothing one can simply take for granted, and that this disclosure relies on the mastery of a specific technique of vision (namely, observation). Like Crary, I am thus in some sense interested in the preconditions of vision. Simple as observation may appear, more often than not we will encounter literary characters who fail, for some reason or other, in the transition from the seeing of images to the seeing of sequences, which is at the core of the observational process. The disclosure of reality through vision thus does depend on the mastering of a technique (i.e. observation), and in highlighting these preconditions of vision, my book stands in a line of scholarship similar to that of Crary, who highlights the rising awareness of the physiological preconditions of vision.

Another important definition of observation has been advanced in the volume *Observation in Science and Literature* (2013), edited by Rüdiger Campe, Jocelyn Holland, and Elisabeth Strowick. Drawing on a tradition of aesthetic theory and philosophy that reaches from Alexander Gottlieb Baumgarten's *Philosophische Briefe* [Philosophical Letters] (1741) to Hegel's *Phänomenologie des Geistes* [Phenomenology of Spirit] (1807), Campe, Holland, and Strowick differentiate between *description* as a first order registration of things, and *observation* as a second order reflection on the process of description.[14] "Observation comes into its own as the critique and undoing of [...] description" (Campe, Holland, Strowick 2013, 371), the editors state in the preface:

> [B]eing an observer in the sense of Baumgarten, Goethe, or Hegel, the subject has a reflective, and therefore active, relationship to its own perceptual receptivity. (Campe, Holland, Strowick 2013, 373).

I concur with this analysis of observation as an activity that presupposes and extends description. However, in contrast to Campe, Holland, and Strowick, I study observation not as a reflection of description; instead, I look at observation as a practice that gives an account of how an initially described state of being develops over time. In observation, two ways of seeing are combined: a seeing of static images (description) and a seeing of dynamic sequences (narration). In my definition, observation aims at the understanding of the object of description (which is now followed over time, so that it reveals 'its story'), not at the understanding of the activity of description itself. Observation, in other words, is not a second-order activity; observation, as I define it, is a complex first-order activity that plays a role, in different ways, in both science and literature.

With the notion of observation as a second-order activity, we have arrived also at the work of Niklas Luhmann. Luhmann has likely had the greatest influence on discussions of the concept of observation, at least in the German academic discourse, of the last several decades. Famously, Luhmann distinguishes between first-order observation and second-order observation (the observation of observation, in other words), and Luhmann argues that second-order observation is a defining feature of modernity. Be it in academic research, in the economy, or in politics, modern societal processes are crucially based on the systematic observation of others' observations (Luhmann 1997, 105–08). Academic research, for instance, as the present literature review plainly illustrates, functions through the examination of others' examination of a specific content

---

14 This interpretation of observation as a second order procedure is in line, moreover, with the results of Christoph Hoffmann's 2006 study of scientific procedures of observation in the late eighteenth and nineteenth centuries. According to Hoffmann, the emerging regimes of observation were characterized precisely by the attention not only to the ultimate object of scrutiny, but also to the sensual apparatus that is used in the observation: "Enger gefaßt betrifft ein Regime der Beobachtung die Verwicklung des Beobachters und der verwendeten Instrumente in die Untersuchung des jeweiligen Forschungsgegenstands." (Hoffmann 2006, 14) [In a narrower sense, an observational regime concerns the entanglement of the observer and the used instruments into the examination of the respective objects of scrutiny. – Unless otherwise noted, translations in this book are my own.]

area. And in the economy, we essentially gauge how much we will pay for a given good by observing how others price this same object. In his 1995 book *Die Kunst der Gesellschaft* [The Art of Society], Luhmann calls such practices of second-order observation in modernity an "evolutionär höchst unwahrscheinlicher und heute zugleich [...] ganz normaler Tatbestand" (Luhmann 1997, 105) [an evolutionarily most unlikely and at the same time a completely normal fact in today's world]. But Luhmann is important not only for having brought our attention to the ubiquity of second-order processes in modernity, but also for his striking definition of first-order observation. Luhmann proposes that observation always consists in an implicit distinction to the end of pointing to only one side of this distinction. Observation thus includes a twofold activity of distinction and signification (*Unterscheidung* and *Bezeichnung*), but in this process only one side of the distinction (the one signified) is actually actively known to the observer. Observation, in other words, always also produces a blind spot – and while second-order observation can become aware of this blind spot, it too will necessarily produce another such blind spot.

Fascinating and influential as Luhmann's theory of observation (first- and second-order) is, it has little bearing on the present study. That being said, one could certainly think of ways in which at the very least Luhmann's theory of second-order observation is related to literature. The volume by Campe, Holland, and Strowick points in that direction. Moreover, Andreas Kablitz's theory of "Realism as a Poetics of Observation" (to cite the title of his 2003 essay) also deserves attention in this context. Kablitz wants to move away from an understanding of realism that is too heavily focused on focalization, i. e. the fact that we are presented with the world of the novel through the eyes of a character inhabiting this world. Instead, Kablitz stresses the ways in which the omniscient narrator works alongside the characters of the story in a structure of universal observation. In his discussion of Balzac, Kablitz explicitly refers to Luhmann to suggest that the relation of the narrator to that of the characters is that of a second-order observer: "To use Niklas Luhmann's words, the narrator observes observations and is able to uncover the generalities behind them for that very reason" (Kablitz 2003, 126).

There are, in other words, meaningful connections between Luhmann's definition of observation and a theory of literary realism. My own line of inquiry, however, takes me in a different direction. Instead of focusing on the relation between different observers (characters and narrators, first- and second-order observers), I am interested in the ways in which one individual observer has to navigate between two different forms of seeing – a seeing of images and a seeing of sequences.

The chapters of this book

Before delving, for the rest of this book, into individual readings of texts that help us analyze the distinct steps in the procedure of observation against the background of a range of different literary and cultural contexts, I will devote one chapter (chapter 1) to a brief survey of the historical debates on description and narration in literary theory. It is against the backdrop of this survey, which expands on the broad outline provided here in the introduction, that we can fully appreciate the significance of literary observation as a combination of description and narration in the following chapters.

My readings in these subsequent chapters engage with novels, novellas, and short stories by Alain-René Lesage, Nicolas Edme Rétif de la Bretonne, Johann Wolfgang Goethe, Georg Büchner, Edgar Allan Poe, and Arthur Conan Doyle. This selection reflects the fact that the development of eighteenth- and nineteenth-century narrative was a genuinely pan-European (or even transatlantic) endeavor and is best studied by looking at more than one national literature. Whether consciously or not (but likely more often consciously than not), writers from many European countries worked through a problem that they shared with others: the problem of how to construct a complete visual reality (which includes static images and dynamic sequences) in narrative texts.

Equally important as the geographical expanse of my study is its wide historical scope, ranging from the early Enlightenment to the eve of Modernism. The two centuries between Alain-René Lesage and Arthur Conan Doyle are united by the dominance of observation as a scientific method[15] as well as by the development of the form of the modern realist novel. As the present study will show, the common narrow definition of the realist novel as the product of the nineteenth century alone precludes us from seeing important lines of development that already begin in the early eighteenth century.

At the core of the following interpretations of French, British, and German fiction stand three narrative models from the beginning, middle, and end of the period that I study. The beginning is marked by Alain-René Lesage's novel *Le Diable boiteux* (1707, revised version 1726). This novel, which I analyze in

---

15 In the second half of the nineteenth century, however, the dominance of observation was increasingly questioned as the practice of experiment gained importance (Daston 2008; see also my comments above as well as chapter 3). Moreover, by saying that observation reigned supreme in the eighteenth and nineteenth centuries, I do not imply that the observational procedures did not change between 1700 and 1900. Lorraine Daston and Peter Galison describe the differences between eighteenth- and nineteenth-century practices of observation in great detail in their study *Objectivity* (2007).

the second chapter, tells the story of a devil who takes a student on a flying tour over the city of Madrid, removes the roofs of the houses, and explains the sequences of events leading up to the presently visible scenes. Lesage's novel, which remained very popular for almost two hundred years, until it was forgotten virtually overnight at the end of the nineteenth century, is an important reference for many later stories about observation, including texts by E.T.A. Hoffmann and Arthur Conan Doyle. But the aerial spectatorship in *Le Diable boiteux* stops short of being observation proper. The premise of this novel is that the presently visible tableaux of urban life in Madrid are always in need of stories to explain them, and these stories have to be provided by the omniscient devil; they cannot be directly observed. With important exceptions, the visual appears in Lesage's novel only in the form of describable, momentary *tableaux*, which the novel's protagonists see under the roofs of the houses in Madrid. The visible and describable tableaux serve as the starting point for the stories that the devil narrates, but they do not contain these stories.

This limitation of the visual is lifted in Rétif de la Bretonne's monumental novel *Les Nuits de Paris* (1788), which I study as the second main narrative model (chapter 3). Rétif's novel, which is composed of the reports of the narrator's encounters on his nocturnal walks through the French capital, distinguishes in its numerous episodes carefully between the different stages of the procedure in which a story is disclosed through observation. This ideal procedure starts with the initial phase of spectatorship, in which the narrator is open to being struck by any curious tableau, and it continues with the focused, prolonged recording of this tableau that captures its changes over time. Observation proper is dependent both on the initial openness to be struck by anything new, which remains limited to the seeing of images, and on the subsequent period of sustained, focused watching, which alone can capture a sequence.

Among all the texts discussed in this book, Rétif's novel offers us the most successful version of observation. As I show through exemplary readings of texts by Goethe, Büchner, and Poe (all in chapter 4), the ideal procedure of observation on which Rétif's novel relies is the implicit target of a range of complications in the decades around 1800.[16]

As suggestive as the model of observation in Rétif's *Les Nuits de Paris* is, it fails to recognize the possibility that the initially perceived, describable image itself contains legible traces of a sequence of events. Precisely this expectation,

---

16 In some way, however, Rétif's novel itself already contains an important complication of the procedure of observation. The point is not so much that the observations in Rétif's novel fail, but that Rétif's novel explicitly raises the question of how the successful observations at the heart of this novel relate to the framework of storytelling in which they are embedded.

however, defines many emerging nineteenth-century scientific observational endeavors, most prominently the geological reconstruction of the long history of the earth from currently visible traces in the landscape. This way of superimposing image and sequence to create a complex temporality of observation finds its literary counterpart in Watson's account of the detective work of Sherlock Holmes – and this is the third and final narrative model that I analyze in this book (chapter 5). Sherlock Holmes, according to Watson, observes the history of the crime in the crime scene. Image and sequence, description and narration collapse into one in Holmes's observations. As such, Holmes's analyses bring the literary history of observation that began with Lesage to an end. Whereas the narration of stories remains wholly divorced from the description of the visible tableaux in *Le diable boiteux*, description and narration tend to coincide in Sherlock Holmes's detective work. Not coincidentally, one might venture to say, Doyle's detective series is also the last prominent literary work that directly references *Le Diable boiteux*. One could not think of a text that deviates more strongly from the narratological premises of Lesage's novel while still sharing its fundamental interest in the visual.

However, a closer look at Doyle's stories reveals that Holmes's analyses often rely less on the patient observation of the individual cases at hand, than on a classificatory calculus that works through the recognition that the present case is in important ways identical to a series of previous cases. To be sure, Holmes does see minute details – but only in order to link the cases through these details to already existing cases. In this sense, Holmes merely classifies; he is not a true observer because he is not open to finding genuinely new knowledge in an extended process of observation. Depending on the perspective, Sherlock Holmes is thus either the perfect observer or not an observer at all. On the one hand, Holmes overcomes all the limitations that defined earlier procedures and collapses the different phases of the observational process into one; but on the other hand, he also tends to replace observation with mere classification. While the Sherlock Holmes stories perform for the reader observations in the sense that newly appearing images are so thoroughly described that their own dynamic history emerges from them, these same stories also call our attention to the fact that this performance of observation remains dependent on a fundamental misrepresentation: Sherlock Holmes himself does not observe; he does not navigate between the openness to seeing new objects appear and the focused attention on a single object; instead, he systematically scans his environment according to a pre-existing system of classification. In this peculiar tension between Watson's portrayal of Sherlock Holmes for the reader on the one hand, and Sherlock Holmes's actual visual practice on the other hand, the era of literary observation reaches its end in a double sense: as both its ultimate goal and point of total decline.

My reading of *Sherlock Holmes* aims to revise some of the key claims of Carlo Ginzburg's seminal article "Morelli, Freud and Sherlock Holmes: Clues and Scientific Method" (1980). According to Ginzburg, Sherlock Holmes's observations should be understood as an example of the new nineteenth-century interest in the individual. However, as I show through a critical re-reading of the different historical examples that Ginzburg lists alongside *Sherlock Holmes*, behind many of the seemingly individualizing procedures lurks a classificatory rationale. While I do not deny the importance of individualizing knowledge in the nineteenth century, I argue for a more nuanced understanding of the extent to which different procedures of the era truly deal with individuals.

One final disclaimer: the sequence of the three narrative models in this book – Lesage, Rétif, Doyle – is meant only partly as a historical argument; to some extent, it is also just a means of organization. My argument is historical not only in the sense that, broadly speaking, the development of narrative strategies of observation unfolds contemporary to the development of scientific observational procedures from the late seventeenth century to the late nineteenth century. I also suggest that, first, Lesage's early-eighteenth-century novel of aerial spectatorship was for many later writers an important source of inspiration; second, that such a detailed account of the mechanics of observations as we find around 1800 in Rétif's *Les Nuits de Paris* most likely cannot be found in novels from a century earlier; and, third, that procedures such as Holmes's are – despite the important precursor in Voltaire's *Zadig* (1747) – essentially new in the literature of the late nineteenth century. But all of this is not to say that there is a strict chain of causation or development leading from Lesage through Rétif to Doyle, or that each episode in this literary history directly corresponds to a specific stage in the history of knowledge. Nor do I even claim that the chosen model texts are immediately and in every aspect representative of their respective period. The realist literature of the eighteenth and nineteenth centuries is certainly much richer and more diverse than any three texts could indicate. Finally, we certainly continue to see observations after 1900. Like countless other aspects of literary realism (including free indirect discourse, descriptive detail, and the attention to social factors, to name just a few), the technique of literary observation remains an important feature of storytelling after the nineteenth century. Some further reflections on the historicity of observation and on observation's afterlife in the twentieth and twenty-first centuries can be found in the conclusion. For all these reasons, therefore, the division of this book into three main narrative models from the early eighteenth to the late nineteenth centuries should be understood largely as a heuristic aid to distinguish between certain fundamental possibilities of storytelling and to construct in the form of a successive development the paradigmatic solution that observation brings to the problematic relation between description and narration, which has haunted literary realism and its theories for a long time.

Last but not least, it bears repeating that the readings in this book are not meant to capture the complete history of literary observation – tracing the first appearance of this technique, as it were, and its changes and popularity through the ages. A book that wanted to tell such a history (and this would be an interesting study indeed), would have to patiently identify, collect, and analyze all those countless little scenes of observation that are similar to the one from the beginning of *Madame of Bovary* with which I opened this book. Instead of writing such a full *history* of literary observation, the present study offers only a few sketches toward a cultural *narratology* of literary observation – highlighting important formal features of literary observation and pointing to the ways in which cases of failure and success of observation emerge from particular sets of literary and cultural preconditions.

Chapter 1: Description and Narration

One reason why literary observation is important is that it bridges the divide between descriptive and narrative passages in a text. Observation moves from the description of a suddenly appearing image of a person (or object) to the narration of how this person (or object) develops over time. It relies on the possibility of description to interrupt the narrative flow with a vivid image, and it sets this static image of description in motion. Observation carries the visuality of description into the narrative and thus functions as a reality effect: evoking a world that is real not just in its visuality (a product of description), but also in its dynamic development (a product of narration). But before we can fruitfully analyze the procedure of literary observation, we should clarify, historically and theoretically, what is at stake in the terms description and narration. The three main questions that I will ask in this chapter are, first, what are (some of) the dominant historical functions of description; second, how can description be said to visualize its objects; and, third, what is the relation between description, narration, and literary realism?

The functions of description

As I suggested in the introduction, description has long been associated with visualization. Descriptions do not propel the plot: they draw attention to the particular, to how things look (to how things *are*, independent of anything things *do*). But it is important to recognize here also that visualization is by no means the only function of description. When, for instance, early modern historiography demands that the narration of events be enriched with the description of the geographical circumstances under which the events occurred, this has less to do with the rhetorical strategy of visualization, and more with an emerging understanding that the narrated actions in some sense depend on the environment in which they are undertaken (that the events are only possible because of their specific environment). Description, in other words, does not only visualize, but it also contextualizes, and we have to understand both of these functions as well as how these functions relate to each other.

The German preface to Giovanni Botero's treatise *Le relationi universali* [Universal Accounts] (1591)[1] is a remarkable early historiographical text in which the

---

1 The German translation from 1596 is entitled *Allgemeine Weltbeschreibung* [Universal Description of the World].

https://doi.org/10.1515/9783110594348-042

call for the contextualization of the narrated events through description explicitly appears.[2] This preface starts out, conventionally enough, by insisting on the necessity to learn from the examples of history – a trope familiar since Cicero under the rubric *historia magistra vitae* [history is the teacher of life] (see Koselleck 2004, 26 – 42); and it is to Cicero himself that the author of the preface actually points. The "weise Heide Cicero" (Botero 1596, without pages) [wise heathen Cicero], we are told, already knew that "nescire quid ante se natum acciderit, id est semper esse puerum" (Botero 1596, without pages) [not to know what happened before you were born is to remain always a child]. But shortly after this familiar trope, the author tells us rather abruptly that the communication of examples from history does not consist only in telling what happened, but also in describing where and how it happened:

> Es gehöret [...] zu einer Historia nicht allein dieses/ daß man etwas erzehle/ was sich begeben hab/ sondern muß auch wissen wo und mit was gelegenheit sich ein jede Sach habe zugetragen/ vnd dernach auff die *descriptiones locorum, urbium,* &c. besondern achtung muß gegeben werden. (Botero 1596, without pages; see Campe 2002, 261)

> It is fitting for a history not only to tell what happened, but we must also know where and under what circumstances any given event took place; and particular attention must thereby be devoted to the *descriptiones locorum, urbium & c.* (Campe 2012, 237)

We are not offered a complete explanation as to why we actually need to supplement the narration of events with the description of the places where these events took place. But it is immediately clear that this claim is crucial to the preface as Botero's entire work is, as the German title announces, an "allgemeine Weltbeschreibung" [universal description of the world]. And we can at least deduce – *ex negativo* – that the function of the demanded "*descriptiones locorum, urbium & c*" is not visualization. Visualization, the preface makes clear a moment later, is accomplished much more directly by the engravings in the volume, "durch welche ein jeder orth dem Leser gleichsam *in re praesenti* vor Augen gestellt wird" (Botero 1596, without pages) [through which each location is brought *in re praesenti* in front of the reader's eyes]. To bring something before the eyes of the reader, as if the things were present (*in re praesenti*): this would otherwise be the function of description. But the author of the preface to Botero's book does not appear to think about description in these terms. Instead, description is necessary to understand and profit from the historical examples: to understand these examples we must know where the examples are located.

---

2 It is through the German translation that Botero's text, which presents one of the first demographic studies, is carried over into the discourse of historiography (Campe 2002, 261).

In combining the traditional assumption according to which one can learn from the examples of history on the one hand with an arguably much more modern insistence that these examples have to be precisely located through description on the other hand, the preface to Botero's work is a curious transitional text. For the insistence on the necessity of contextualizing description is, in some sense, in tension with the assumption that history offers us examples that we can easily apply to the present condition. As the German historian Reinhart Koselleck has shown, in the two centuries after Botero's work (and notably around 1800), the old episteme of history as a pool of examples that can easily be applied to the present was increasingly replaced by an understanding of history as a development from a fundamentally different past into an unknown and open future. When history is no longer thought to repeat itself, we cannot easily learn from its examples anymore.[3] The preface to Botero's work, by insisting that the examples of history are in some way (we are not really told why) specific to the context in which they occurred, already makes a first step in the direction of the new concept of history: the idea that any event depends on its particular environment (which is rendered in description) lends itself much more readily to the idea that history is a forward-movement without direct repetitions. Conversely, if we assumed that history does repeat itself and that we can therefore easily learn from examples, we are less likely to place great importance on the specific context of a given event. To be sure, the author of the preface to Botero's work does not realize the tension between his traditional reliance on the trope that we can learn from history and his new insistence on the importance of description. But this tension is there; and what historically prevailed in the centuries after the publication of Botero's work – albeit not without considerable exceptions and trends to the contrary[4] – is the emphasis on description, and with that, the new concept of history as progress.

I dwell here on the preface to Botero's book because it exemplifies, in a nutshell, one important discourse on description. This is a discourse that sees in de-

---

3 Koselleck develops this argument, for instance, in his essays "Historia Magistra Vitae: The Dissolution of the Topos into the Perspective of a Modernized Historical Process" (Koselleck 2004, 26–42) and "Historical Criteria of the Modern Concept of Revolution" (Koselleck 2004, 43–75).
4 In her essay "Description by Omission" (2005), Lorraine Daston analyzes an actual decrease in descriptive detail in scientific writing from the seventeenth to the eighteenth centuries, and she links this development to the emerging sense of a universal (human) nature that was thought to prevail over all local specificities: "The metric system (allegedly nature's own measure), the 'Droits de l'Homme' (proclaimed as 'natural, imprescriptible, and inalienable rights'), and the Napoleonic wars that attempted to export both of them from Milan to Moscow were ultimately all anchored in the universality of nature." (Daston 2005, 23)

scription the reflection of an emerging understanding of how human actions are shaped by their environment. In this study, I focus largely on a different aspect of description, namely the idea that description makes things visible. But it is also important to recognize that these two distinct ways of looking at description are not entirely alien to each other. The realist attempt to visualize the world of the novel through description can be seen to be motivated to some extent by an attempt to make the grounding of the narrated actions in their environment *evident*. These descriptions literally *show* the materiality of the world from which the actions emerge. It will therefore also not always be fruitful to distinguish entirely between these two aspects of description (as either visualizing or grounding in a material context). These two aspects offer distinct and yet overlapping and mutually supporting arguments for description.

Over the centuries following the publication of Botero's work, the emphasis on "the *descriptiones locorum, urbium & c*" spread far beyond the confines of historiographical narrative. As Rüdiger Campe argues, the demand in modern historiography for the combination of the narration of events with the description of the setting shaped particularly the development of the modern novel (Campe 2002, 261). In fact, the novel adopted description much more readily than did historiography, which always countered the demand for description with an equally strong imperative not to include any unnecessary detail. As Cynthia Sundberg Wall has shown, there is a long line of "[c]riticism from Horace and Aristotle to René Rapin (1621–87), Peter Whalley (1722–91), and of course, Samuel Johnson (1706–84) [that] catechized against ornament and superfluity [...] in history writing" (Wall 2006, 221). Although eighteenth-century novelists and critics were divided over the question of description as well, the skeptical voices had here only limited impact and the amount of descriptive detail steadily increased in the literary works of the eighteenth century.[5]

While the urge to describe became steadily more prominent in the eighteenth century, it gained special momentum in nineteenth-century novels by writ-

---

5 See Watt 1957 and, more recently, Wall 2006 and Delon 2009. Wall links the rise of description in the eighteenth century to the developing consumer culture and the resulting increased interest in 'things.' For an exploration of the ways in which description developed in eighteenth-century science and literature, see Stalnaker 2010; on the development of scientific description from the fifteenth to the eighteenth centuries, see Ogilvie 2006. For a broader exploration of the cultural forces that favored or inhibited description in the eighteenth century, see also John Bender and Michael Marrinan's volume *Regimes of Description: In the Archive of the Eighteenth Century*. Erich Auerbach's *Mimesis* (1946) remains an important survey of the multifold changes that the style of literary description underwent in the Western tradition from Homer to Virgina Woolf. For examples of criticisms of descriptive detail, see also Hamon 1991, 7, 27, 145.

ers including Dickens, Flaubert, Stifter, and Zola.[6] Given this prevalence of description in novels of the period that we first associate with literary realism – i.e. the nineteenth century – it is not surprising that realism and description are sometimes almost regarded as synonymous.

It is interesting to note in this context that Émile Zola, likely one the most prolific writers of description in the nineteenth century, still justifies description in a way that recalls the justification in the preface to Botero's historiographical treatise (while also developing this justification in much more detail and in a language that borrows from the new scientific discourses of the nineteenth century). Similar to the preface to Botero's work, Zola declares in his programmatic essay "De la description" [On Description] (1880) that humans and their actions cannot be separated from the setting in which they occur: "Nous estimons que l'homme ne peut être séparé de son milieu" (Hamon 1991, 157) [We believe that man cannot be separated from his milieu]. In Zola's understanding, literature can fulfill its potential and become "scientifique" (Hamon 1991, 155) [scientific] only if it thoroughly describes the social background from which each protagonist emerges.

But despite this testimony by one of realism's foremost writers, description's ability to explain the background from which characters emerge is arguably not the most important factor for the close association between description and realism. Rather, description seems to be especially apt to fulfill a realist poetics because it has the potential to directly evoke the materiality of the world as it visually appears to us. Narration tells us what happens in the world; description tells what the world looks like. Description makes us see the world. And seeing is the sense that is, according to the modern hierarchy of senses that was codified in numerous treatises of the sixteenth and seventeenth centuries, most intimately linked to the perception of the world 'as it really is' (see Adler 1989, 11). As the sociologist Judith Adler points out in this context, it is the "eyewitness" and not the "hearsay" that is judged in modernity as "legally admissible evidence and ground for valid judgment" (Adler 1989, 11). Or, as Francis Bacon states in

---

5 Svetlana Alpers argues that a similar turn from narration to description can be observed in the realist movements in painting in the seventeenth and nineteenth centuries: "this singular combination of an attention to imitation or description with a suspension of narrative action is not an isolated feature of some works by Caravaggio. It is also a characteristic of some of the greatest works of the leading seventeenth-century realist painters—Velázquez, Rembrandt, and Vermeer. Further, this phenomenon seems not to be limited to the seventeenth century, for it reappears once more in French realist art of the nineteenth century—in Courbet and Manet." (Alpers 1976, 15)

his *Novum Organum* [The New Organon] (1620), one of the founding texts of modern empiricism, "Inter sensus autem manifestum est partes primas tenere Visum quoad informationem" (Bacon 1889, 491) [it is evident that sight holds first place among the senses as far as information is concerned (Bacon 2000, 171)]. Bacon is important to cite here, as the modern empiricism that he helps prepare the way for reconfigures truth as a product of sense perception – and especially of vision. To run the risk of oversimplification,[7] we might say that, ever since Bacon, "true" is what is accessible to the senses (to sight) – not what is guaranteed by ancient authority or syllogistic reasoning alone. If we want to convey truth after the scientific revolution of early modernity, we have to convey sense perception and especially vision – and in literature (as elsewhere), this becomes the task of description.

We all know, of course, that sense perception, and especially vision, can be subjective and deceptive and that evidence taken from vision alone is not always trustworthy. This knowledge is in no way new. Indeed, the deceptiveness and ambiguity of vision has been a major theme in the European arts and sciences at least since early modernity, if not since antiquity. As Martin Jay has amply shown, even during those periods that he identifies most strongly with what he calls "ocularcentrism" – Ancient Greek culture, the scientific revolution, the Enlightenment, and the nineteenth century – we persistently also find various musings on the shortcomings of visual evidence (Jay 1993, 21–209). Bacon himself, who in some way inaugurates the authority of visual evidence in the early modern sciences, is entirely aware of the defects of human sight. But this awareness never leads him to question the principal link between vision and reality. The strong overall criticism of our reliance on vision to disclose reality is, as is already revealed by the title of Jay's book – *Downcast Eyes: The Denigration of Vision in Twentieth-Century French Thought* (1993) – a quite recent phenomenon.

This longstanding sympathy between visuality and reality is also what Peter Brooks captures in *Realist Vision* (2005), a study of the nineteenth-century novel. Brooks defines the realist literature from Balzac to Zola chiefly through its "visuality: its primary attention to the visible world, the observation and representation of persons and things" (Brooks 2005, 72). Reality, for Brooks, is linked to vision, and vision is captured through description. In some sense, the argument about the relation between description and vision may appear weak. Why

---

7 A more nuanced overview of the legacy of "ocularcentrism" in European culture until its demise around 1900 can be found in the first three chapters of Martin Jay's 1993 study *Downcast Eyes: The Denigration of Vision in Twentieth-Century French Thought* (Jay 1993, 21–209).

would description allow us primarily to *see* the world? If description represents static objects, we might just as well say that description allows us to smell, taste, or touch the world – and, in fact, description does sometimes achieve these things as well.[8] In all of these different ways, description shows what things are like – independent of any action that they perform. The only reason that there should be a privileged link between description and vision is that if we perceive reality primarily visually and if it were the task of description to capture reality, we should then expect description to focus on vision. This does not appear to be a strong link – and yet historical convention has fully established this link between description and vision.[9] For the association between description and visuality reaches back as far as Quintilian's classical rhetorical treatise *Institutio Oratoria*. In his discussion of the virtues of vivid description (*enargeia*), Quintilian asserts: "Magna virtus est res de quibus loquimur clare atque, ut cerni videantur, enuntiare" (Quintilian 1976, vol. 3 244) [It is a great gift to be able to set forth the facts on which we are speaking clearly and vividly (*videantur*; literally, so that they may be seen). (Quintilian 1976, vol. 3, 245)] Quintilian strongly influences a long list of rhetoricians and writers from late antiquity through the nineteenth century who describe the function of description precisely in its ability to make us see things (see Hamon 1991, 8, 23, 30 et al.). The French *Encyclopédie*, for instance, as Cynthia Sundberg Wall also points out, asserts that "description" is "a figure of thought by development of which, instead of simply indicating an object, makes it somehow visible" (Wall 2006, 11; Wall's translation).[10] And in the novels by Radcliffe and Scott, we are presented, Wall claims, "with a fully visualized *setting* in which events will occur. We are given the visual world [...]." (Wall 2006, 5, emphasis in the original)

---

8 This is not to say that it might not be easier for description to capture visual images than to capture smell or taste. Read, for instance, the text on a wine bottle and judge whether you are more likely to form a picture of the region that is being described or sense the bouquet that is being described. But this difference seems to have to do with experience and convention and not with any inherent limitation of description.

9 Beth Innocenti notes in her study of descriptions from the Roman rhetorical tradition that descriptions that appeal to the senses of smell and touch remain rare. She argues: "Because sight was held to be the most vivid sense and because the sight of scenes may more often promote the response for which the speaker aims than, for example, the sounds and smells of a scene, the recurrent features of vivid descriptions tend to include primarily visual descriptions." (Innocenti 1994, 370–371)

10 The entry in the *Enclyclopédie* is by Nicolas Beauzée. The opening paragraph reads: "La Description est une figure de pensée par dévelopement, qui, au lieu d'indiquer simplement un objet, le rend en quelque sorte visible, par l'exposition vive et animée des propriétés et des circonstances les plus intéressantes." (Hamon 1991, 211, emphasis in the original)

In highlighting this long tradition of linking description with vision, we should not lose sight of the heterogeneity of this tradition. Quintilian, for instance, develops his theory of descriptions in a rhetorical context, for people speaking in a law court before judges. One of the main functions of visualization here is to assist in persuasion (Innocenti 1994, 363). Quintilian writes in the passage of his *Institutio Oratoria* in which he deals with description:

> Non enim satis efficit neque, ut debet, plene dominatur oratio, si usque ad aures valet atque ea sibi iudex, de quibus cognoscit, narrari credit, non exprimi et oculis mentis ostendi. (Quintilian 1976, vol. 3, 244)

> For oratory fails of its full effect and does not assert itself as it should, if its appeal is merely to the hearing, and if the judge merely feels that the facts on which he has to give his decision are being narrated to him, and not displayed in their living truth to the eyes of his mind. (Quintilian 1976, vol. 3, 245)

This element of persuasion is largely absent in the nineteenth-century novel, in which visualization (through description) serves to show us the real material world from which the narrative emerges. If description in the novel wants to persuade us of anything all, it is the 'reality' of the described world.

What do we see when we read a description?

The heterogeneity of the theories of description does not only affect the question of the purpose of visualization, but also the question of what precisely we are thought to see when reading a description. The two extreme positions in this respect are, first, that through description one sees the object that is being described and to the extent that it is being described (and everyone reading the description sees the same thing), and, second, that the actual description only initiates in the recipient a process of visualization in which the recipient will see more (and different) things than literally expressed in the description (and different recipients could produce different images). While cursory statements on description's power to present things as if we saw them before our eyes sometimes suggest the first of these options, any more detailed account of description would probably have to include at least some elements of the second position. Let us here briefly consider two examples, one by Quintilian and one by Flaubert. Both of them lean to the section position, but the different ways in which they do so are worth notice.

To be sure, when Quintilian praises a description of two boxers (by Aeneas), saying that its details "nobis illam pugilum congredientium faciem ita osten-dun:, ut non clarior futura fuerit spectantibus" (Quintilian 1976, vol. 3, 246) [give us such a picture of the two boxers confronting each other for the fight, that it could not have been clearer had we been actual spectators (Quintilian 1976, vol. 3, 247)] it may sound as if Quintilian were siding with the first position: through the detailed description we are given a complete picture of the boxers. But already with the next example, taken from Cicero's *Verres* [The Verrine Orations] (first century BCE), it becomes clear that Quintilian actually thinks a description effective only if it allows us to see *more* than what is being directly expressed:

> An quisquam tam procul a concipiendis imaginibus rerum abest, ut non, cum illa in Verrem legit, *Stetit soleatus praetor populi Romani cum pallio purpureo tunicaque talari muliercula nixus in litore*, non solum ipsos intueri videatur et locum et habitum, sed quaedam etiam ex iis, quae dicta non sunt, sibi ipse adstruat? Ego certe mihi cernere videor et vultum et oculos et deformes utriusque blanditias et eorum qui aderant tacitam aversationem ac timidam verecundiam. (Quintilian 1976, vol. 3, 246, emphasis in the original)

> Is there anybody so incapable of forming a mental picture of a scene that, when he reads the following passage from the Verrines, he does not seem not merely to see the actors in the scene, the place itself and their very dress, but even to imagine to himself other details that the orator does not describe? "There on the shore stood the praetor, the representative of the Roman people, with slippered feet, robed in a purple cloak, a tunic streaming to his heels, and leaning on the arm of this worthless woman." For my own part, I seem to see before my eyes his face, his eyes, the unseemly blandishments of himself and his paramour, the silent loathing and frightened shame of those who viewed the scene. (Quintilian 1976, vol. 3, 247)

What Quintilian here professes to see indeed markedly exceeds anything captured in the actual description that he cites.[11] For Quintilian, the point does not seem to be that readers subjectively supplement the information provided in the description, and that each reader supplements the image in a different way (even though this possibility is not directly excluded). All readers might actually see the same thing, and yet the viewed image still goes far beyond what is being explicitly said. Quintilian's idea that description makes us see, in a possibly quite determined way, something beyond the description – showing very specifically something that is *not* mentioned – is interestingly echoed by the fact that the first stylistic technique that Quintilian discusses after vivid description (*enargeia*) is the simile. Similes are a means of illustration that work precisely by

---

11 The description offered by Cicero and the supplement by Quintilian are, of course, quite different – with an account of clothing on the one hand and a focus on emotions on the other hand.

such a detour, showing us something that is not described by describing instead something else.

Quintilian, in sum, is fully aware of the fact that the image produced in the recipient is not directly expressed in the words of the description. What is, however, as yet only of little importance in Quintilian is the idea that what is being described could lead to different images in different recipients. Precisely this idea is of foremost importance to Flaubert. Flaubert, however, goes beyond the relatively simple fact of reception that different readers will form a different picture of what is being described in a novel. His much more profound claim is that the effect of visualization fundamentally depends on this ambiguity in the text. A good description, in other words, is not one that captures the object in the most precise terms, but one that leaves the most room to the readers to realize the image in whichever way their specific memories and inclinations would lead them to do. Flaubert was therefore also vehemently opposed to any illustrations in his novels. The point for him is not that these illustrations render the text incorrectly, but that they counteract the text's strategy to allow each and every reader to visualize the description in his or her own way (and then to exclaim tautologically: yes, I have seen that before!). In a letter to Ernest Duplan from 12 June 1862, Flaubert explains:

> Jamais, moi vivant, on ne me m'illustrera, parce-que : la plus belle description littéraire est dévorée par le plus piètre dessin. Du moment qu'un type est fixé par le crayon, il perd ce caractère de généralité, cette concordance avec mille objets connus qui font dire au lecteur : «J'ai vu cela» ou « Cela doit être». Une femme dessinée ressemble à une femme, voilà tout. L'idée est dès lors fermée, complète, et toutes les phrases sont inutiles, tandis qu'une femme écrite fait rêver à mille femmes. (Hamon 1991, 152)

> Never in my life shall I be illustrated, because the most beautiful literary description is destroyed by the most miserable drawing. From the moment that the character is fixed by the pencil, this character loses its generality, this agreement with a thousand known objects that allow the reader to say: "I have seen that" or "It must be like that." A woman who is drawn resembles one woman, and that is all. The idea is from then on closed, complete, and all the phrases are useless, while a woman who exists in writing makes us dream of a thousand women.

The example that Flaubert chooses here – the illustration or description of a woman – is telling. The visualizing power of description relies, in Flaubert's view, on the desire of each and every reader: in some limited sense, it is really the reader's desire that is visualized through the (sufficiently general) description, not the described object itself.

Flaubert's brief statement is remarkable for its Copernican shift, so to speak, in which the image produced by description now seems to illustrate almost more the interiority of the reader than the object described. But this statement also complicates a common trope about realism as a literary style concerned with detail and with the particular, because it links the ability of description to visualize to its "caractère de généralité." However, Flaubert, who is surely aware that his own descriptions (think of the description of the schoolboy Charles Bovary with which I opened this book) are by no means simply generic or unspecific, evidently does not see a contradiction here. Although pointing to individual details, his descriptions are nevertheless general in the sense that the details leave room for a variety of concrete images to be completed by each reader.

The mechanisms of visualization are curious, and, as the examples from Quintilian and Flaubert indicate, there is a considerable range in the explanations of visualization through description. But why does the precise mechanism of visualization matter to the present study? Indeed, the distinction between the two perspectives on visualization that I outline here (that description either fully determines an image or that it merely initiates a process of visualization that depends much on the reader) might not be very important for the argument that I develop in this book. Observation, I argue, carries the visuality of static description over into narrative – but how precisely this process of visualization happens is of minor importance. And yet it is useful to develop these different perspectives on the mechanism of visualization in some detail – not only to raise awareness for the complexity of the process of visualization on which I rely in my argument, but also to show the range of concepts of description and visualization that can be aligned with the project of this book. Saying this, of course, also implies that there are ways of understanding description that fall outside the scope of this book and this too should briefly be recognized here.

It is fruitful for this purpose to contrast Flaubert's resistance against illustration with the resistance Franz Kafka expressed some fifty years later concerning any illustration of the insect at the heart of his novella *Die Verwandlung* [Metamorphosis] (1915). In a letter to his publisher Kurt Wolff (dated 25 October 1915), Kafka urges: "Das Insekt selbst kann nicht gezeichnet werden" (Kafka 1966, 136) [The insect itself cannot be drawn]. The two statements by Flaubert and Kafka are much further apart than any initial reading may lead one to assume (after all, one might say, both Flaubert and Kafka are in some way skeptical of illustration). In contrast to Kafka, Flaubert admits that illustration – say, of Madame Bovary – is possible. Such an illustration would realize one possible view of this character based on the descriptions found in the book. Illustration may not be desirable (because it impedes the process of visualization for the readers), but it is possible. For Kafka, by contrast, the point seems to be that any illustration would essentially obscure the fact that visualization is impossible – that the exterior reality remains hidden, both to the protagonists of his texts and to the reader. The insect into which Gregor Samsa is transformed at the outset of Kafka's story cannot be pictured – neither mentally, nor graphically (in print). Kafka's claim that illustration (and visualization) is impossible is reflected in his approach to description in this text. Kafka's poetics of description are a poetics of the undepictable – refusing the possibility of visualization to which description (naturally) strives. To be sure, the text seems to begin, familiarly enough (recall again the opening of *Madame Bovary*), with the sudden introduction of a striking image – quite in the tradition of what I call in this book the initial moment of a literary observation. One morning, Gregor Samsa wakes up and finds himself transformed into an insect:

> Als Gregor Samsa eines Morgens aus unruhigen Träumen erwachte, fand er sich in seinem Bett zu einem ungeheuren Ungeziefer verwandelt. Er lag auf seinem panzerartig harten Rücken und sah, wenn er den Kopf ein wenig hob, seinen gewölbten, braunen, von bogenförmigen Versteifungen geteilten Bauch, auf dessen Höhe sich die Bettdecke, zum gänzlichen Niedergleiten bereit, kaum noch erhalten konnte. Seine vielen, im Vergleich zu seinem sonstigen Umfang kläglich dünnen Beine flimmerten ihm hilflos vor den Augen. (Kafka 2002, 96)

> As Gregor Samsa awoke one morning from uneasy dreams he found himself transformed in his bed into a gigantic insect. He was lying on his hard, as it were armor-plated, back and when he lifted his head a little he could see his domelike brown belly divided into stiff arched segments on top of which the bed quilt could hardly keep in position and was about to slide off completely. His numerous legs, which were pitifully thin compared to the rest of his bulk, waved [literally "glimmered" (*flimmerten*)] helplessly before his eyes. (Kafka 1993, 75)

What appears at first reading to be a rather precise description of the insect is in fact shot through with markers of the impossibility to visualize what is being described – impossible for the protagonist through whose eyes we are presented with the world of this story, but also for the readers. The numerous legs of the insect "flimmertern" [glimmered] before the protagonist's eyes, thus complicating any clear vision of this insect, which is, in any case, largely still covered by the bed quilt. Indeed, this fragile bed quilt, precariously resting on the insect's stomach and about slip off, is arguably the most remarkable feature of this short description in which the central object itself (the insect) remains impossible to picture. What Kafka constructs here in his description of the insect is indeed a kind of 'non-being,' an "*un*geheure[s] *Un*geziefer" (my emphasis): the English translation "gigantic insect" does not fully capture the negativity implied by the double-use of the negative prefix "un-" in the German original. While Kafka offers a description of the insect that is at the heart of his story, he counteracts any attempt to form a vivid image of this insect.

Years later, Kafka opens his novel *Das Schloß* [The Castle] with a description in which the impossibility to visualize what is being described receives even stronger emphasis. The protagonist here gazes at a castle that remains entirely hidden in fog and darkness:

> Es war spät abends als K. ankam. Das Dorf lag in tiefem Schnee. Vom Schloßberg war nichts zu sehen, Nebel und Finsternis umgaben ihn, auch nicht der schwächste Lichtschein deutete das Schloß an. Lange stand K. auf der Holzbrücke die von der Landstraße zum Dorf führt und blickte in die scheinbare Leere empor. (Kafka 1994, 9)

> It was late when K. arrived. The village lay under deep snow. There was no sign of the Castle hill, fog and darkness surrounded it, not even the faintest gleam of light suggested the large Castle. K. stood a long time on the wooden bridge that leads from the main road to the village, gazing upward into the seeming emptiness. (Kafka 1998, 1)

Be it the in the figure of the "ungeheures Ungeziefer" at the outset of *Die Verwandlung* or that of the invisible castle (at which K. nevertheless stares for a long time) at the beginning of *Das Schloß*, Kafka's concern is with an undoing of the visualizing function of description. We may still somehow believe that we see something (an insect and a castle) – but any careful reading of the text would tell us that what we see here is not really grounded in the description itself. While the description points us to an object, it also insists on the impossibility of fully picturing this object.

I cite Kafka's texts here so extensively because they mark a point in literary history –Modernism – at which the fundamental preconditions of literary observation are called into question. Of course, Kafka shows only one facet of Modernism. Where Kafka stresses in his poetics of description the limits of human perception to grasp an exterior reality, other writers of that period – we may think of Virgina Woolf – simply shift their interest away from the exterior reality as such to its reflection in the protagonists of their texts.[12] But in one way or another, the idea of description as a tool to evoke visual images of an exterior reality is sidelined. This is not to say that earlier centuries show us uncomplicated procedures of observation. Quite to the contrary, most of the readings in this book will highlight just how precarious the seemingly so simple transition from the description of an image to the narration of a story, which is at the heart of the observational process, appeared to eighteenth- and nineteenth-century writers. But the basic underlying assumptions that the world is fundamentally visually accessible (that we can see it) and that description can, however imprecisely or undeterminedly (as in the case of Flaubert), capture the world in its visual appearance,

---

12 See my brief discussion of Virginia Woolf in the conclusion.

are rarely called into question. Modernism – exemplified here by the works of Franz Kafka – departs from a century-long belief in the potential of description to present us with exterior objects as if they were before our eyes. This is one major reason why the onset of Modernism also marks the historical endpoint of the present study. To be sure, one will easily find literary observations in the literature of the twentieth and twenty-first century – but largely (albeit not exclusively) only insofar as other stylistic elements of older times also live on after the rupture of Modernism: as an old exercise that is no longer fully representative of its contemporaneous culture.[13]

Description, narration, and literary realism

But let us leave Kafka's modernism for now as a distant shadow on the horizon and go back to the literature of the late eighteenth and nineteenth centuries (the period of literary realism), for which Peter Brooks and Cynthia Sundberg Wall rightly emphasize the rich, visualizing descriptions of rural and urban landscapes, houses, furniture, clothes, and bodies. However, even if it is the case, as most eighteenth- and nineteenth-century writers assumed, that descriptions provide us with a visual image of the world (however imprecise this image may be), the question remains whether description alone can meaningfully indicate reality. It is this question, as well as the range of answers that modern critics have provided to it, on which I focus in the remaining pages of this chapter.

The demand for description that we saw in the preface to Botero's work of historiography implied that description should accompany narration, not replace it. The function of description is to give an account of the circumstances that allow us to situate the narrated actions and events in the world. But what are these circumstances without a narrative in them? Even if we say that description allows us to see the world – what is this visible world if it remains in the frozen state of description, void of any temporal development? To be sure, the descriptions in a text imply an understanding of the deep connection between human action and the world in which this action occurs. By themselves, however, these descriptions – or so it seems – do not fully show reality. My point here is perhaps not so much that description is, as Gérard Genette would have it, "*ancilla narrationis*" (Genette 1982, 134) [the handmaiden of narration], but that it is

---

13 On Modernism's undoing of description, see extensively Heinz J. Drügh's study *Ästhetik der Beschreibung* [Aesthetics of Description] (2006). See also my remarks on Modernism and the afterlife of literary observation in the conclusion.

the *partner* of narration in the portrayal of a dynamic visual reality – and that description thus still relies on narration.

No one has discussed these shortcomings of mere description in more extreme terms than Georg Lukács. In his 1936 essay "Erzählen oder Beschreiben?" Lukács attacks leading realist and naturalist writers of the nineteenth century, including Flaubert, Zola, and Stifter, for their heavy reliance on description and their inability to create strong narratives in which descriptions could occupy a meaningful place. Instead of constructing such integrative narratives, these writers offer in their texts "Bilder, die [...] so unverbunden nebeneinander hängen wie Bilder in einem Museum" (Lukács 1955, 125) [pictures (...) as isolated and unrelated to each other as pictures in a museum (Lukács 1970, 134)].[14] According to Lukács, only narrative with its temporal extension and its ability to depict human action can give a coherent account of (social) reality:

> Wo und wie wird aber diese Wahrheit sichtbar? Es ist nicht nur für die Wissenschaft, nicht nur für die wissenschaftlich fundierte Politik, sondern auch für die praktische Menschenkenntnis im Alltagsleben klar, daß diese Wahrheit des Lebens sich nur in der Praxis des Menschen offenbaren kann, in seinen Taten und Handlungen. (Lukács 1955, 115)

> Where and how is this truth revealed? It is clear not only in science and in politics founded on a scientific basis but also in man's everyday practical common sense that truth is revealed only in practice, in deeds and actions. (Lukács 1970, 123)

In contrast to narration, description produces images and is thus, by definition, unable to render actions and relations of cause and effect. Description turns the world into a still life, in which things exist simply side-by-side. It cannot answer the question of how the individual elements affect and transform one another:

> Die Beschreibung gibt also keine wirkliche Poesie der Dinge, verwandelt aber die Menschen in Zustände, in Bestandteile von Stilleben. Die Eigenschaften der Menschen existieren nebeneinander und werden in diesem Nebeneinander beschrieben, statt wechselseitig einander zu durchdringen und damit die lebendige Einheit der Persönlichkeit in ihren ver-

---

14 One may hear in this formulation a distant echo of André Breton's criticism of descriptions from the period of realism as mere "postcards." In the *Manifeste du surréalisme*, Breton writes: "Et les descriptions ! Rien n'est comparable au néant de celles-ci ; ce n'est que superpositions d'images de catalogue, l'auteur en prend de plus en plus à son aise, il saisit l'occasion de me glisser ses cartes postales, il cherche à me faire tomber d'accord avec lui sur des lieux communs [...]" (Hamon 1991, 176) [And the descriptions! There is nothing to which their vacuity can be compared; they are nothing but so many superimposed images taken from some stock catalogue, which the author utilizes more and more whenever he chooses; he seizes the opportunity to slip me his postcards, he tries to make me agree with him about the clichés. (Breton 1969, 7)]

schiedenartigsten Äußerungen, in ihren widerspruchvollsten Handlungen zu bezeugen. (Lukács 1955, 129)

Description provides no true poetry of things but transforms people into conditions, into components of still lives. In description, men's qualities exist side by side and are so represented; they do not interpenetrate or reciprocally affect each other so as to reveal the vital unity of personality within varied manifestations and amidst contradictory actions. (Lukács 1970, 139)

The Marxist thinker Lukács detects behind the nineteenth-century realist writers' inability or unwillingness to produce narrative a deeply corrupt bourgeois ideology that is detached from the actual social conditions of life. Enjoying a simultaneously relatively stable, comfortable, and (seemingly) powerless position, the bourgeois writer becomes blind to the actions and transformations that occur every day. The bourgeois writer loses sight of the actually occurring actions and changes in society and thus betrays the task of literature to depict "d[ie] Bewährung oder d[a]s Versage[n] der menschlichen Absichten in der Praxis" (Lukács 1955, 115) [the success or failure of human purpose in the test of practice (Lukács 1970, 124)]. The consequence of this unduly complacent bourgeois realism is not only a literature that is so boring that many readers seek refuge in trivial fiction, which still provides narrative and action, but also that literature loses sight of what reality really is – human action under social constraints:

Man sieht, daß der moderne Realismus, daß die Methode der Beobachtung und der Beschreibung durch den Verlust der Fähigkeit, die wirkliche Bewegung des Lebensvorgangs zu gestalten, die kapitalistische Wirklichkeit abgeschwächt und verkleinlicht, unangemessen widerspiegelt. (Lukács 1955, 136–137)

For all its close observation and description, modern realism has lost its capacity to depict the dynamics of life, and thus its representation of capitalist reality is inadequate, diluted and constrained. (Lukács 1970, 147)

Lukács's provocative claim is that a good portion of nineteenth-century realism is, contrary to what its name promises, singularly unable to portray reality. According to Lukács, literary realism remains detached from reality as long as it only describes characters and places as they appear in a given moment. Lukács demands that the realist writer, instead of merely conjuring up static descriptions, has to show how these described objects and people develop and gain their meaning as parts of temporal sequences and human interactions. What is real, for Lukács, is only the coherence of narrative and action, not the still life of description. For Lukács, the descriptions of objects and circumstances can only ever be justified if the described object plays an important role in the human actions that form the narrative. Lukács distinguishes, somewhat polemically between the "tiefpoetische Wirkung der aus dem Schiffbruch zusammengesuchten Instrumente im 'Robinson'" (Lukács 1955, 127) [profound poetic effect of the tools rescued from the shipwreck in *Robinson Crusoe* (Lukács 1970, 136)] and the "überflüssig" (Lukács 1955, 127) [superfluous (Lukács 1970, 136)] character of "eine[r] beliebige[n] Beschreibung aus Zola" (Lukács 1955, 127) [any description at all in Zola (Lukács 1970, 136)].

Lukács's argument evokes, to some extent, the much older criticism of description in Lessing's *Laokoon* essays from 1766 – and Lukács does indeed draw on Lessing in support of his argument.[15] But while Lessing also argues against the reliance on description in narrative texts, his main concern is, in contrast to Lukács, the effect of description on the reader, not the relation between description and reality.[16] For Lessing, description may very well be ontologically realistic (it can represent what the world is like); it just does not evoke a very clear and strong picture of reality (this is a concern that Lessing generally shared with some art critics and even natural historians of his time, who also cautioned against an overflow of description in their specific disciplines; see Stalnaker 2010, 13 – 15). Lessing argues that due to the fact that texts develop in temporal succession (they are told over time), they are most apt to create the illusion of temporal sequences, not of stable states of being, where things exist next to each other (this is done much better in painting).[17] Discussing the description of a beautiful woman in a poem by Ariosto, Lessing admits that this description may accurately *represent* female beauty; his concern, however, is that no reader will be able to form a *vivid picture* of the described woman. Contrasting his po-

---

15 See Lukács 1955, 128 – 129. For an extensive discussion of the position of Lessing's *Laokoon* within the context of Enlightenment aesthetics, see Wellbery 1984. See also the interpretative overview and survey of the *Laokoon* scholarship by Fick 2000, 216 – 241.

16 It may not be amiss to recall, moreover, that Lukács and Lessing have very different literatures in mind when they condemn description. While Lukács turns against some of the today still most prominent novelists of nineteenth-century realism and naturalism, including Stifter, Flaubert, and Zola, Lessing attacks today largely forgotten eighteenth-century descriptive poems, like Ewald von Kleist's *Der Frühling* [Spring] (1749) and Albrecht von Haller's *Die Alpen* [The Alps] (1729).

17 With the emphasis on the limitations of specific media, Lessing's argument differs from the dominant eighteenth-century argument against the abundance of description in other disciplines. As Joanna Stalnaker explains, the more prominent claim by French scholars, such as the naturalist Louis-Jean-Marie Daubenton or the art critic and writer Denis Diderot, was that description, in order to visualize objects, needs to omit some detail. Paradoxically, it is easier to form a vivid picture of a described object if one is given less information (Stalnaker 2010, 13 – 15; see also Daston 2005). This argument is, at best, a simplification of Lessing's argument.

sition to that of the critic Dolce, who admires Ariosto's description, Lessing notes:

> Dolce bewundert darin die Kenntnisse, welche der Dichter von der körperlichen Schönheit zu haben zeigt; ich aber sehe bloß auf die Wirkung, welche diese Kenntnisse, in Worte ausgedrückt, auf meine Einbildungskraft haben können. (Lessing 1965, 131)

> Dolce admires the knowledge of physical beauty which the poet displays in them [Ariosto's verses], while I am concerned only with the effect which this knowledge, when expressed in words, has on my imagination. (Lessing 1984, 106 – 107)

Lessing's claim is that description can *represent* reality; however, it cannot effectively *evoke* it (see also Beaujour 1981, 40).[18] This claim is different from the one that Lukács makes. Lukács denies description – at least in those cases in which description does not directly serve the narrative – any potential to portray reality.

However, even if one generally concurs with Lukács's (and Lessing's) insistence on the superiority of narration, it is possible to challenge Lukács's representation of the development of eighteenth- and nineteenth-century fiction as a misled journey to description alone. While the increase in isolated descriptive passages and the relative neglect of narration certainly captures one prominent line of development, I reconstruct in this book a technique of literary realism that strives to extend the visuality that is conventionally associated merely with description to the narrative. Literary observations set the visual world of description in motion; and it is through this setting-in-motion of visual description that literary texts achieve an important *reality effect*. Observations appear 'real' in that they perform for the reader a real process of perception, which moves from the seeing of a curious image to its sustained recording over time.

---

18 Nevertheless, Lessing does not seem to be completely consistent in his argument. At times, he suggests that literary narration is indeed better equipped to represent reality. Discussing the depiction of gods in painting and poetry, Lessing claims that in painting the gods appear only as personifications of virtues, whereas narrative shows the gods as "real beings:"

„Bei dem Künstler sind sie personifizierte Abstrakta, die beständig die nämliche Charakterisierung behalten müssen, wenn sie erkenntlich sein sollen. Bei dem Dichter hingegen sind sie *wirkliche handelnde Wesen*, die über ihren allgemeinen Charakter noch andere Eigenschaften und Affekten haben, welche nach Gelegenheit der Umstände vor jenen vorstechen können." (Lessing 1965, 73, my emphasis)

"To the artist they [gods and spiritual beings] are personified abstractions which must always retain the same characteristics if they are to be recognized. To the poet, on the other hand, they are *real, acting beings* who, in addition to their general character, possess other qualities and feelings which, as circumstances demand, may stand out more prominently than the former." (Lessing 1984, 52, my emphasis)

My claim about the integration of description and narration in literary procedures of observation differs significantly from the work of other scholars who have tried to safeguard description's place in a poetics of literary realism against Lukács's criticisms. Among these competing accounts of the relevance of description, Roland Barthes's much-discussed 1968 essay "L'effet de réel" [The Reality Effect] stands out, from which I borrow the term reality effect (while defining this reality effect in very different ways). In this essay, Barthes begins by outlining a similar dichotomy as Lukács between narration, which creates coherence and meaning, and description, which interrupts this coherence. In contrast to Lukács, however, Barthes assigns the descriptive passages, which are not integrated into the narrative, the potential to signify the real. Description has this potential, Barthes argues, precisely *because* it does not contribute to the narrative context, which creates coherence and meaning in the texts. The more out of context and meaningless the described details are, the better they designate the real. These details are purely real in sense that they are simply *there* – for themselves as it were, and not motivated by any (narrative) function. Barthes's idea of a reality of insignificant, decontextualized details – which, incidentally, strongly echoes eighteenth-century accounts of descriptive detail by Denis Diderot and, slightly later, by Charles Nodier (see Hamon 1991, 97–101) – has to be understood in the context of Barthes's general understanding of reality in this essay. When Barthes speaks in his essay of the reality effect of the insignificant details of description, he does not suggest that these details actually signify the concrete, extra-textual real world in any meaningful way. Quite to the contrary, the function of these details is only to indicate the category of "the real" as such, without regard to its actual concrete elements. Referring to his two examples from Flaubert and Michelet, Barthes explains:

> [C]ar dans le moment même où ces détails sont réputés dénoter directement le réel, ils ne font rien d'autre, sans le dire, que le signifier : le baromètre de Flaubert, la petite porte de Michelet ne disent finalement rien d'autre que ceci : *nous sommes le réel* ; c'est la catégorie du «réel» (et non ses contenus contingents) qui est alors signifiée [...]. (Barthes 1968, 88, emphasis in the original)

> Just when these details are reputed to denote the real directly, all that they do—without saying so—is *signify* it; Flaubert's barometer, Michelet's little door finally say nothing but this: *we are the real*; it is the category of "the real" (and not its contingent contents) which is then signified. (Barthes 1986, 148, emphasis in the original)

According to Barthes, the reality effect thus points to a reality that is utterly void of any concrete content. The reality indicated by the descriptive details in nineteenth-century fiction is, is in other words, an empty category. In this critical assertion about description, Barthes seems to concur with Lukács. However, in contrast to Lukács, Barthes also suggests that the empty reality that is evoked by the descriptive detail in realist fiction is representative of our general conception of reality in modernity. The reality of modernity is, according to Barthes, a meaningless fetish – an entity that is perpetually evoked as an almost magical value in and for itself and that is never substantiated or contextualized. It does not matter *what* the real is and how it functions; the real owes all its authority to being there despite and outside of all context and function.[19] We moderns, Barthes suggests, venerate the real as tourists and as guests of historical exhibitions, lining up to see monuments whose concrete historical importance is of no interest to us. We simply want to see the 'real thing,' no matter wherein its significance lies:

> [I]l est logique que le réalisme littéraire ait été, à quelques décennies près, contemporain du règne de l'histoire «objective» à quoi il faut ajouter le développement actuel des techniques, des œuvres et des institutions fondées sur le besoin incessant d'authentifier le «réel» : la photographie (témoin brut de «ce qui a été là»), le reportage, les expositions d'objets anciens (le succès du show Toutankhamon le montre assez), le tourisme des monuments et des lieux historiques. Tout cela dit que le «réel» est réputé se suffire à lui-même, qu'il est assez fort pour démentir toute idée de «fonction», que son énonciation n'a nul besoin d'être intégrée dans une structure et que l'*avoir-été-là* des choses est un principe suffisant de la parole. (Barthes 1968, 87–88, emphasis in the original)

> [I]t is logical that literary realism should have been—give or take a few decades—contemporary with the regnum of "objective" history, to which must be added the contemporary development of techniques, of works, and institutions, based on the incessant need to authenticate the "real": the photograph (immediate witness of "what was here"), reportage, exhibitions of ancient objects (the success of the Tutankhamen show makes this quite clear), the tourism of monuments and historical sites. All this shows that the "real" is supposed to be self-sufficient, that it is strong enough to belie any notion of "function," that its "speech-act" has no need to be integrated into a structure and that the *having-been-there* of things is a sufficient principle of speech. (Barthes 1986, 146–147, emphasis in the original).

While Barthes argues that other historical periods may have denoted reality differently and may have seen it rather on the side of meaning and coherence (Barthes 1986, 147), he does not immediately value one of these concepts of reality higher than the other; he is not interested in any exploration of the true form of reality. There is, for Barthes, nothing more to say about reality beyond the different ways in which texts and cultures construct their sense of the real.

---

19 Barthes constructs a similar argument in his 1980 essay *La chambre claire* [Camera Lucida], in which he discusses the effect some photographs have on him. This effect is not caused by the photographs' general meaning-producing structure, which Barthes calls the "studium"; instead the photographs' effect is caused by precisely those points in the picture that interfere with the structure – Barthes calls these points the "punctum" (Barthes 1981, 25–27).

This distinguishes him from Lukács. While Barthes remains skeptical of nineteenth-century realist attempts to denote the real, he does not argue from the perspective of a strong independent concept of the real.

Some critics have since tried to advance the argument that there is in fact a more substantive reality behind the decontextualized descriptive details that Barthes analyzes as the reality effect. In his inaugural lecture at the Humboldt University in Berlin, the German literary critic Klaus Scherpe, for instance, called description a crucial tool for the representation of the modern human condition, precisely because of description's potential to disrupt the deceptive coherence of narrative (Scherpe 1996). Where narrative creates a potentially false promise of coherence and meaning, description focuses on the moment, disrupts the sequence, and can thus point to the fragility of any totalizing construction. However, Scherpe's praise of the disruptive force of description in the modernist writings of Marcel Proust, Alfred Döblin, and others, falls in a certain way behind Lukács. What Scherpe depicts as radically modernist descriptions that reflect the crisis of meaning in modern times might be reinterpreted – in Lukács's terms – as a bourgeois misunderstanding of modernity.[20]

As different as the theoretical positions of Lukács, Barthes, and Scherpe are, not one of them questions the principal dichotomy of static description and dynamic and coherent narration. To be perfectly clear, I, too, believe that it makes sense to keep description and narration as two fundamental forms of writing clearly apart from one another. However, I argue that there exists with observation a procedure that importantly combines description with narration and that offers a solution to the long-standing tension between the realism of visualizing description and the realism of narration. To be sure, we will not be able to find this procedure of observation in every single text of the eighteenth and nineteenth centuries. But we do see this procedure regularly, in a wide range of texts. And while these observations may take up only a fraction of the novels and stories in which they appear, they do fulfill the function of an important reality effect. They signify a reality different from that outlined by Barthes. It is not simply a reality of insignificant details; it is a reality that is concrete in its descriptive visuality and rich in its narrative temporal dynamism. Observations perform the perceptual process in which a curious image appears and is set in motion, thus mimicking the way we encounter and study new aspects of reality.

---

20 Scherpe avoids an all-too direct confrontation with Lukács by focusing on modernist writers of the early twentieth century – instead of on nineteenth-century realist and naturalist writers, as Lukács does. The structure of the argument, however, remains similar, regardless of whether one speaks about modernity in general or Modernism in particular.

In recent years, critics have increasingly questioned the clear-cut binary between static description and dynamic narration, which Lukács, Barthes, and Scherpe claim to find in many eighteenth-, nineteenth-, and early twentieth-century literary texts. Pointing to the difficulty to ascertain the distinction between description and narration in any given passage, Ruth Ronen, for example, advocated the complete abolition of this distinction from the critical vocabulary (Ronen 1997). Much less radically, Gérard Genette – himself a proponent of the usage of the binary – conceded that a purely narrative text without any descriptive aspects is impossible (Genette 1982, 134; see also Schmid 2010, 5–6).

This skepticism regarding the clear-cut binary of description and narration is especially striking in the case of the Stifter scholarship of the last decades. Like the work of few other nineteenth-century European writers, Adalbert Stifter's novels and novellas have been traditionally associated with long descriptions – in particular descriptions of the rural landscapes of the Austrian Empire, where most of Stifter's novels and novellas are set. "Love of detail" and "very little action" (Heller 1891, iii) are, for instance, the main attributes that the nineteenth-century critic Otto Heller uses in a short portrait of Stifter.[21] The emphasis on description and its inherent static character in Stifter's works is sometimes understood as a political message. In this reading, the apparently unchanging landscapes that are constructed in Stifter's descriptions signify the – real or illusory – idyll and natural safe haven beyond all threatening narratives of social change in the big cities. Some newer scholarship, however, questions whether Stifter's descriptions of the rural landscapes are really as static as they appeared to generations of readers. Peter Schnyder, for instance, suggests in a recent interpretation of Stifter's novel *Der Nachsommer* [The Indian Summer] that Stifter, in accordance with contemporary geological research, does not merely offer static images in his descriptions of the Austrian Alps. Instead, these descriptions contain the knowledge of the dynamic, albeit slow, historical processes in which the present landscapes have been shaped. For Schnyder, Stifter's description of the mountain landscapes *is* narration and does depict change (Schnyder 2009).[22]

However, as interesting as Schnyder's reading of Stifter is, it is, I argue, not useful to completely obfuscate the opposition between narration and descrip-

---

21 For a more extensive discussion of description in Stifter, see, e. g., Stern 1964.
22 Tove Holmes's interpretation of *Der Nachsommer* also works at the undoing of the traditionally assumed opposition of description and narration in Stifter's works. For her, the narrative arch is constructed in the sequence of increasingly rich and complex descriptions. These descriptions, Holmes argues, track and display the interior growth of the protagonist through whose eyes we encounter the described objects and landscapes. The narrative, in other words, consists of a series of descriptions (Holmes 2010).

tion, which Lukács analyzes so forcefully in nineteenth-century realist literature. Lukács provides with his strong distinction between description and narration a valuable conceptual vocabulary (which is not the same as to say that it is always easy to apply this vocabulary unequivocally to an individual literary text), and we may even follow his claim that descriptive passages have to be integrated into a coherent narrative in order to create an apt model of reality. As Lukács explains, these descriptive elements can gain their significance only through their relation to the overarching narrative. This is the case because reality is, for Lukács, most fundamentally temporally extended human action under social constraints.

However, Lukács's account of nineteenth-century realist literature still remains deficient in that Lukács possibly underestimates the degree to which the problematic relation between narration and description is already a central concern in some of the writers whom he attacks most harshly, such as Stifter (see Lukács 1955, 122–123; on the relation between Stifter and Lukács see also Geulen 1998). Stifter's possibly best-known novella, "Bergkristall" [Rock Crystal] (1845), which tells the story of two children who get lost in a snowstorm, is a case in point. This story interestingly complicates Lukács's image of Stifter's descriptive realism.

"Bergkristall," I argue, can be read as a clever commentary on the problem of aligning narration and description. Like many of Stifter's stories, "Bergkristall" begins with a long description of the setting – in this case a remote village and its surrounding valley and mountains. Indeed, the initial description takes up about a fifth of the entire text. Surprisingly, however, this minute description proves almost useless for the ensuing narrative about the two children in the snowstorm. This snowstorm makes any orientation in the previously described landscape impossible. The surveyed landscape of the initial description and the obfuscated landscape that appears during the narrative of the snowstorm are incompatible. The initial description does little to set the scene for the later story. Rather, the story undermines the promise of description to meaningfully show us – and orient us in – a material, visual reality.

In the story, Konrad, who is the older of the two siblings, keeps trying to locate himself in the snowstorm based on the information that has been introduced to the reader of the story in the initial description. This information, however, has become meaningless because the snow has destroyed all previous markers. Stifter brilliantly shows the boy's inability to think beyond the initial description. At the same time that the boy admits that he can see neither path, nor trees, nor a prominent monument ("Die Unglückssäule"; in the English translation "the post") – all of which are familiar to the reader from the initial description –, the boy keeps referring to path, trees, and monument as means of orientation:

> „Ich weiß es nicht", antwortete der Knabe, „ich kann heute die Bäume nicht sehen und den Weg nicht erkennen, weil er so weiß ist. Die Unglücksäule werden wir wohl gar nicht sehen, weil so viel Schnee liegen wird, daß sie verhüllt sein wird und daß kaum ein Gräschen oder ein Arm des schwarzen Kreuzes hervorragen wird. Aber es macht nichts. Wir gehen immer auf dem Wege fort, der Weg geht zwischen den Bäumen, und wenn er zu dem Platze der Unglückssäule kommt, dann wird er abwärtsgehen, wir gehen auf ihm fort, und wenn er aus den Bäumen hinausgeht, dann sind wir schon auf den Wiesen von Gschaid, dann kömmt der Steg, und dann haben wir nicht mehr weit nach Hause." (Stifter 1962–1972, vol. 4, 211)

> "I don't know," answered the brother. "This time, I can't make out the trees, or the road because it is so white. We may not see the post at all, because there is so much snow it will be covered up, and hardly a grass-blade or arm of the cross will stick out. But that's nothing. We'll just keep straight on; the road leads through the trees and when it gets to the place where the post is, then it will start downhill and we keep right on it and when it comes out of the woods we are in Gschaid meadows; then comes the footbridge, and we're not far from home." (Stifter 2008, 39–40)

Description and narration remain completely divorced from each other in "Bergkristall." But this separation of description and narration is not so much a weakness of the composition, as Lukács would have it, as it is itself the very topic of Stifter's novella.

However, Lukács ignores not only such implicit commentaries on the relation between description and narration in the works of the writers whom he dismisses wholeheartedly; more importantly, he does not pay enough attention to the possible ways of literary storytelling to combine the contrary strategies of detailed, visualizing description and coherent narration. The main argument of this book is that there was in fact an important procedure in narrative texts from the early enlightenment era to the eve of Modernism that helped to combine description and narration. With this I am not referring merely to the various ways in which nineteenth-century writers tried to create seamless transitions between the descriptive passages, which evoke the world of the novel in its visibility, and the narrative passages, which propel the plot. To be sure, such techniques certainly exist, and Philippe Hamon has analyzed them in detail (Hamon 1981). As Hamon shows, writers regularly invented miniature fictions that allowed the descriptive passages to appear as an organic part of the narrative: the description of a room, for instance, starts as the protagonist is looking around, waiting for another person to appear; the description of the neighborhood is inserted as the protagonist looks out of the window or walks through the streets. While these and similar 'tricks' allow the description indeed to blend in with the narrative, there still remains a fundamental functional difference between the evocation of visuality through description and the creation of a coherent plot through narration. The procedures that I am interested in, in contrast, highlight how the act of seeing is itself constructed through the combination of description and narration. These procedures are characterized simultaneously by the "primary attention to the visible world" (Brooks 2005, 72) of description, and by the coherent, temporally extended sequences of narration. Observation turns the paintings of description into moving pictures, and thus leads from description to narration. Observation is the answer to the opposed demands of description and narration in modern realist writing. It evokes reality visually, and it propels this visual reality in a dynamic development.

In the remaining chapters of this book, I analyze a range of works in which such procedures of observation take center stage. More precisely, I will pair a few texts in which we see successful observations – mainly, *Les Nuits de Paris* (chapter 3) and, with some qualifications, *Sherlock Holmes* (chapter 5) – with a variety of texts in which the crucial transition from static image to moving image (from description to narration) fails. Through this kaleidoscope of failures, we can appreciate the astonishing literary and cultural-historical preconditions of a rather common technique of realist writing.

Chapter 2: Before Observation
(*Le Diable boiteux*)

I began this book with the opening scene of *Madame Bovary*. There, we saw a classroom door open and a new student appear. The boy is described in some detail and then he is watched over time so that we see him act (albeit in a very limited way) and so that a narrative begins. I call this combination of the description of a static image with the narration of an extended sequence of actions an observation, and I argue that observations are an important feature of realist writing. They evoke through their initial description a visual image, and they carry this visuality over into the narrative. They show us a world that appears 'real' both in its visual sensuality, and its dynamic development. The aim of this book is to analyze in both narratological and cultural-historical terms the factors that enable – or complicate – this seemingly so simple literary procedure.

A first important text to consider in this endeavor in some detail is Alain-René Lesage's novel *Le Diable boiteux* (1707).[1] Lesage's novel tells the story of the devil Asmodée, who takes the student Don Cléofas Léandro Perez Zambullo on a flying tour over the nocturnal city of Madrid. On this tour, the devil makes the roofs of the houses disappear, so that he and the student have an unobstructed view onto the interior scenes. With this prominent gesture of a visual disclosure of the world, *Le Diable boiteux* remained, as I will show in the last part of this chapter, an important reference point for two centuries after its publication. In particular, *Le Diable boiteux* is one of the key texts that nineteenth-century realist writers cited to point to a form of literature that claims to be based on a seeing of the world. Alluding to Lesage's novel, Balzac, for instance, signed several of his texts as the "Diable à Paris."

Lesage's *Le Diable boiteux* is an important paradigm for the emerging realist novel. In contrast to other central early texts of the realist tradition – including Cervantes's *Don Quijote* [Don Quixote] (first volume 1605), Grimmelshausen's *Simplicissimus Teutsch* [Simplicius Simplicissimus] (1668), and Behn's *Oroonoko*

---

1 Like any true historical origin, *Le Diable boiteux* has itself important predecessors. The novel is heavily based on Luis Vélez de Guevara's 1641 novel *El Diablo Cojuelo* [The Lame Devil]. This earlier novel already contains the motif of the removal of the roofs, which is central to *Le Diable boiteux*. However, Lesage's novel had a much greater success than Vélez de Guevara's. At least in the German, French, and British traditions of the eighteenth and nineteenth centuries, the scene of the removal of the roofs is essentially associated with Lesage's *Le Diable boiteux* (Saint-Amour 2011, 227).

https://doi.org/10.1515/9783110594348-067

(1683) – it makes the connection between reality and vision one of its key themes.[2] And it was as a text about vision that *Le Diable boiteux* was remembered in the eighteenth and nineteenth centuries. This makes Lesage's novel an important text for my reconstruction of literary observations. What is, however, so very curious about the great legacy of *Le Diable boiteux* for the tradition of a 'visual realism' in general, and for my theory of observation more specifically, is that the visual is actually judged very ambivalently in this novel. To be sure, Lesage's novel stresses a moment of seeing as the beginning of all storytelling – but also only as the beginning. Vision, in *Le Diable boiteux*, remains limited to the seeing of the describable situations that the narratives set out to explain. The scenes or *tableaux* under the roofs of the houses serve as the starting point for the stories that are told by the devil, but, importantly, these stories themselves cannot be gleaned from the visible tableaux themselves. The explicit visuality of the description is rarely carried over into narration: there are, with very few exceptions, no observations in *Le Diable boiteux*. The devil can tell the stories only because of his superior knowledge of all things past and present. In the novel, these many 'invisible' backstories are modeled in the fashion of the contemporary comedy of manners. In total, the devil produces more than a hundred miniature narratives as well as around a dozen novellas of varying length about greedy usurers, old coquettes, gullible husbands, and lovers of all kinds. The tableaux that devil and student see demarcate the final moment of each of these stories. The novel, in other words, consists in the collection of backstories of that which is presently visible, and it adds to this collection a small frame narrative about the spectatorship of devil and student.

But why is the potential of the visual so limited in Lesage's novel? Why is vision so strictly separated from the narration and only allotted a place in the initial description? Why, in other words, do observations remain so rare in this novel that evidently has a great interest in the visual? To understand this, it is crucial to recall the moral principles of this novel, which are laid out in fairly clear terms in the frame narrative. In this frame narrative, we read the story of how the student Don Cléofas, while fleeing some assassins, enters into an astrologer's attic. Here, he encounters the devil Asmodée, who has been caught by the astrologer and who languishes in one of the many bottles stored in the attic. As-

---

2 Of the other novels mentioned here, *Oroonoko* has the most explicit reference to vision. As the narrator states at the outset, "I myself was an eye-witness, to a great part, of what you will find here set down" (Behn 1992, 75). But the important identification of the narrator as an 'eye-witness' (which recurs, for instance, in Defoe's *A Journal of the Plague Year*, 1722) is inserted largely as a general guarantor of truth, not to indicate that the novel will somehow seek to reproduce a fundamentally visual experience of reality.

modée begs the student to deliver him from his narrow confines and promises him, as a reward, great knowledge of the world. The precise wording of the devil's promise holds the key to the understanding of visuality that shapes this novel: "Je vous apprendrai tout ce que vous voudrez savoir. Je vous instruirai de tout ce qui se passe dans le monde. Je vous découvrirai les défauts des hommes." (Le Sage 1984, 33) [I will teach you everything you would like to know. I will instruct you about everything that happens in the world. I will uncover to you the vices of people.] Reading this striking three-part promise, it is important to note that the initially extremely broad promise to teach anything the student wishes to know is more and more defined, until, in the third formulation, the initial promise is reduced to the revelation of the vices or shortcomings of people ("les défauts des hommes"). What this means is that knowledge, in the world of this novel, is essentially defined as the uncovering of vices. Within this moral episteme of uncovering, the visual attains a curiously ambignous – we might even say, contradictory – status. On the one hand, the gesture of uncovering is an essentially visual metaphor, captured most prominently in the lifting of the roofs of the houses: we are allowed to 'see' below the surface the true life of Madrid's citizens. By the same token, however, vision is also always shown to be deceptive: the truth lies beyond what is visible and tentatively contradicts visual appearance. In this latter sense, the staging of an 'uncovering of vices' works by way of contrasting the visual surface with the true essence. Lesage's novel is thus caught in a strange ambivalence in its functionalization of vision. Vision is both relied upon in the visual metaphor of uncovering (lifting the veil of reality) and rejected in a binary of truthful essence and deceptive appearance.

It is in the context of this marked ambivalence that we should understand the strong contrast between highly visual description (the initial seeing of the images under the roofs of the houses) and the strikingly a-visual narration of the backstories provided by the devil. Being aware of this compromise between an affirmation of the visual (in the metaphor of uncovering) and rejection of the visual (as false appearance) then also helps us solve a difficulty that any interpretation of Lesage's novel faces, namely that this novel combines two literary configurations that are at odds with one another. *Le Diable boiteux* prominently introduces a novel form of narration that performs an engagement with the world as it visually appears to us, and it is also a collection of standardized stories in the style of the comedy of manners, with its canon of stock figures and its gesture of moral critique. We can easily analyze each of these two elements separately: the mechanics of seeing and describing the world as it appears to us on the one hand, and the poetics of the comedy of manners with its well-established stock characters and standard plots on the other hand. But it is not as easy to combine these two aspects. The comedy of manners with its stock characters does not need observers; it already possesses a complete inventory of all characters and stories that make up society. Observation, by contrast, demands that we suspend our presumed knowledge of the world and learn about individual cases as they appear to our eyes.

Not surprisingly, the conflation of a proto-observational framework with the well-rehearsed moral stereotypes of the comedy of manners disturbed some twentieth-century critics. Jean Vic, for instance, laments that the heavy reliance on previous literary works counteracts the idea of observation: "L'imitation [...] s'exerce au détriment de l'observation : Lesage se borne à réunir des masques de comédie." (Vic 1920, 516) [The imitation works to the detriment of observation. Lesage limits himself to the assemblage of the masks of comedy.] Vic certainly captures the tension between the proto-observational framework and the stock characters of the comedy of manners correctly, but this tension should be understood less as a shortcoming of Lesage's novel and more as a logical consequence of the novel's framework of moral critique (of moral discovery) and the ensuing ambivalence toward the visual.

One important task in the interpretation of Lesage's novel then consists in taking the novel's engagement with the visual seriously while at the same time keeping in mind the limited importance that the visual has in *Le Diable boiteux*. By taking the novel's engagement with the visual seriously I mean to say that I deal with the novel as a text that has an implicit theory of how it is that we understand the world by seeing it. Indeed, I will devote the first three sections of this chapter to an analysis of the ways in which Lesage's *Le Diable boiteux* stages the distinct steps of the process in which the spectator's pleasure in a pure seeing – a seeing without any defined objects – is transformed into an epistemic practice by compartmentalizing the field of vision and defining the *tableaux* that could theoretically be the object of detailed description and the starting points of observation. With few exceptions, however, devil and student in *Le Diable boiteux* do not engage in the concentrated and sustained watching of the visual objects that they carve out of the initially confused mass of sensory data: instead of watching the momentary scenes develop into a narrative, they stop at the isolation of describable tableaux, and they supplement these tableaux with narratives that detail what happened before. The few exceptional observations in *Le Diable boiteux* are, nonetheless, important to this novel, and in the fourth section of this chapter, I analyze their context as a way to confirm why it is that observation remains, for most of the novel, impossible. Before closing this chapter, I turn finally to the curious fact of literary and cultural history that the ambivalent stance toward the visual, which I take to be a central feature of *Le Diable boiteux*, was essentially forgotten in the productive reception of Lesage's novel in the eighteenth and nineteenth centuries. Lesage's novel was reinterpreted to become a simple cipher for a literature that focuses on the visual experience of the world. I argue, moreover, that, due to this prevalent interpretation of *Le Diable boiteux*, the popularity of Lesage's novel can serve as an index of a more widespread interest in literary observational procedures. For as long as *Le Diable boiteux* remained popular, literary observations were popular as well.

The pleasure of seeing

Standing on the top of a church tower in Madrid and simply stretching his arm, Lesage's devil Asmodée makes the roofs of the surrounding houses disappear, and, despite the dark of the night, the interior reveals itself to the eyes of the student who accompanies him:

> [I]l ne fit simplement qu'étendre le bras droit, et aussitôt tous les toits disparurent. Alors l'écolier vit, comme en plein midi, l'intérieur des maisons, de même, dit Luis Velez de Guévara, qu'on voit le dedans d'un pâté dont on vient d'ôter la croûte. (Le Sage 1984, 40 – 41)

> [H]e simply extended his right arm, and in an instant, all the roofs of the houses disappeared. So the student saw, as in plain daylight, the interior of the houses, in the same fashion, says Luis Vélez de Guevara, in which one sees the inside of a pâté from which one has just removed the crust.

There is something immediately fascinating about this novel's primal gesture of the removal of the roofs and the lightning of the atmosphere. It promises absolute visibility and speaks to our desire to see and stare at the world – without, moreover, this act of seeing affecting the scene in any way. For despite the daylight that suddenly illuminates the night, no one inside the houses is aware of any change.

This striking gesture of uncovering builds on a long cultural history of diabolic revelations and yet remains, as we shall see, importantly distinct from this tradition by foregrounding the pleasure of pure seeing in an unprecedented way. At the beginning of this tradition stands the story of the temptation of Christ in the desert, as it is told in the gospel according to Matthew. In the third and final scene of the temptation-narrative, the devil leads Jesus on a high mountain and shows him "all the kingdoms of the world, and the glory of them; And saith unto him, All these things I will give thee, if thou wilt fall down and worship me" (Matthew 4, 8 – 9; King James Version). In some way, the act of seeing appears to be of only secondary importance here: Christ is *shown* the world merely in order to indicate what he could *possess*. The temptation, one may say, lies in the possession of the world, not in its beholding. But the reception of this scene in early modernity increasingly emphasizes the attraction of the visible independent of the promise of possession. To my knowledge, the earliest work of art foregrounding to some extent the visual as such in the temptation scene is an altarpiece by the Italian Renaissance artist Duccio di Buoninsegna (1255–1319, see Fig. 1). Interestingly, Duccio's altarpiece is also regarded as an early example of European art that begins to be based on the direct survey of nature.[3]

More prominently than in any painting, however, the tradition of diabolic revelation is continued in the early modern Faust stories. While of relatively minor importance in the original German chapbook *Historia von D. Johann Fausten* [History of Dr. Johann Faustus] (1587), the motif of visual discovery is foregrounded in Christopher Marlowe's 1604 play *Doctor Faustus*. One of the first gifts that Faust is granted by the devil after their pact is a flight from his home in Wertenberg in Germany to the papal palace in Rome. In the play, the flight gives rise to Faust's description of Europe as seen from above – or, more precisely, to a series of tourist snapshots of prominent sights on the continent (see Marlowe 1969, 54). Even in the Faust tradition, however, the idea of the visibility of the world is not yet fully divorced from the idea of the possession of the world. The world that Faust is allowed to see is always still potentially his to possess.

It is only with Lesage's *Le Diable boiteux*, published early on in the age of scientific observation, that the *discovery* of a visible world – and not its *possession* – becomes the central gift of the devil. While Lesage's novel relies on a previous tradition of diabolic visual discovery, Lesage is the first to allow great prominence to pure visibility, freed from the idea of ownership. At the same time, moreover, the devil's gift (of an unrestricted overview of, and visual access to the world) has lost almost all associations with sinfulness and temptation (Reiffers 2013, 258). If the devil allows us to see, there does not seem to be any reason not to look.

The great emphasis on the visual discovery of the world in Lesage's novel is well captured in an engraving by Dubercelle for a 1726 edition of *Le Diable boiteux* (Fig. 2). In this engraving, devil and student are shown standing on a tower in the top left of the image, gazing into the interior of houses whose roofs have

---

3 Moritz Reiffers remarks in relation to Duccio's painting of the temptation scene: "Duccio markiert mit seiner Kunst auch allgemein einen mit Beginn des 14. Jahrhunderts sich in Italien neu entwickelnden Zug zur Naturbeobachtung als Grundlage der Kunst." (Reiffers 2013, 74) [Duccio's art generally marks a new tendency, which was emerging in fourteenth-century Italy, to base art on the observation of nature.] At the same time, Reiffers insists that the visual itself is not yet shown here as a real temptation for Christ, who sends the devil away, but does not seem to be leaving his lookout point (Reiffers 2013, 75).

Figure 1: Duccio di Buoninsegna, *The Temptation of Christ on the Mountain*, 1308–1311, New York, Frick Collection. Copyright The Frick Collection.

been removed. The main difference to Duccio's painting of the temptation scene is at once obvious: in contrast to Duccio's painting, which focuses on the conflict between Christ and Devil, Dubercelle's engraving foregrounds the visibility of the world. Dubercelle does more than merely lift the roofs off the houses. He also makes us see through windows; he removes entire front walls and distorts the image's perspective.[4] All this serves to make us see a maximum number of scenes at once. Nothing is supposed to be kept from the eye of the spectator.

---

4 If one does not want to accept a distortion of the perspective, one is forced to assume a highly awkward course of the street that separates the houses on the left from the houses on the right.

Figure 2: Devil and student looking into Madrid's houses. Engraving by Dubercelle for the 1726 edition of *Le Diable boiteux* (Le Sage 1726, 25). Source gallica.bnf.fr / BnF.

Admittedly, it is hard to make much sense of the scenes that are revealed in Dubercelle's engraving. For what do we really see? There is a man walking in an otherwise empty attic; another man sits in a chair and gazes upwards; a group of men and women are at a lively banquet; a man and a woman are engaged in an apparently intense conversation... The engraving abounds with a multiplicity of disparate material at which we can stare; but the engraving does not show us anything in much detail or in a way that would allow us to analyze and understand what exactly happens in the scenes. In this engraving, the promise that we can see all and everything is more important than anything we can actually see.

Here, as in several other cases that we will encounter over the following chapters, the illustrator possesses a privileged understanding of the organization of the visual in the novel. For the student Don Cléofas really reacts similarly to the way a beholder of Dubercelle's engraving will react. When Don Cléofas is first allowed to look into the houses, he does not look at any one particular scene. Instead, he turns his head in every direction, enjoying the pure pleasure of seeing:

> Le spectacle était trop nouveau pour ne pas attirer son attention tout entière. Il promena sa vue de toutes parts, et la diversité des choses qui l'environnaient eut de quoi occuper longtemps sa curiosité. (Le Sage 1984, 41)

> The spectacle was too new not to attract his [the student's] entire attention. He let his gaze wander everywhere, and the diversity of the things that surrounded him would have occupied his curiosity for a long time.

The student does not look at anything in particular, and yet he is enthralled by the scene. The scene takes entire hold of his attention – without him actually being drawn to any individual object. He looks distractedly everywhere and nowhere to satisfy his erring curiosity. He enjoys a pure 'seeing without objects' – he sees, but he sees 'nothing.'

Consider, for the sake of contrast, the type of *absorbed beholder* that Michael Fried analyzed in paintings of the later decades of the eighteenth century, including Jean-Baptiste Simeon Chardin's famous *Soap Bubbles* (c. 1733/1734, see Fig. 3, see Fried 1988, 50 – 51).[5] Chardin's painting shows a teenaged boy intensely gazing at a soap bubble that he is blowing up with a straw. In the background, we can a see younger child avidly staring at the same bubble. Everything centers on the fragile bubble, which we are invited to watch in the same state of absorption as the two boys in the painting.

There is pleasure in absorptive seeing, just as there is pleasure in the distracted seeing practiced by Lesage's student. And both forms of seeing may be called "pornographic" – in the sense in which Frederic Jameson has called the

---

5 I will have more to say about this absorbed spectator in my interpretation of Goethe's *Die Leiden des jungen Werthers* (see chapter 4). Werther, I argue, embodies the type of absorbed spectator that Fried describes in Chardin's painting.

Figure 3: Jean-Baptiste-Simeon Chardin, *Soap Bubbles*, probably 1733/1734, Washington, D.C., National Gallery of Art. Courtesy National Gallery of Art, Washington.

visual as such "pornographic, which is to say that it [i. e. the visual] has its end in rapt mindless fascination" (Jameson 1992, 1).

Nevertheless, these two forms of seeing are fundamentally different from one another. Lesage's student, despite being captured by the activity of seeing, is by no means absorbed, because he does not have an object in whose contemplation he possibly could be absorbed. Instead, he resembles the child who is first given a remote control for his TV and cannot stop zapping the channels, in a manner that is at the same time fascinated and entirely distracted. Lesage's novel approximates the scene of the child zapping the channels as closely as could be possible 250 years before zapping became a technological and cultural reality.

Pleasure and instruction

While the student enjoys his mindless, distracted 'seeing without objects,' the devil insists that the engagement with the visual shall serve more than merely the student's immediate desire to gaze and stare. Instead, this encounter with the world should also be useful by increasing the student's knowledge of life:

> Seigneur don Cléofas, lui dit le Diable, cette confusion d'objets que vous regardez avec tant de plaisir est, à la vérité, très agréable à contempler. Mais ce n'est qu'un amusement frivole. Il faut que je vous le rende utile ; et, pour vous donner une parfaite connaissance de la vie humaine, je veux vous expliquer ce que font toutes ces personnes que vous voyez. Je vais vous découvrir les motifs de leurs actions et vous révéler jusqu'à leurs plus secrètes pensées. (Le Sage 1984, 41)

> Seigneur don Cléofas, said the devil to him, this confusion of objects that you regard with such pleasure is, to be sure, very nice to contemplate. However, it is but a frivolous amusement. I have to render it useful to you; and to give you perfect knowledge of human life, I want to explain to you what all the people whom you see are doing. I will discover the motives of their actions and reveal to you even their most secret thoughts.

The devil's reminder that the pleasure of seeing has to be accompanied by some beneficial insights evokes the ideal of Horatian poetics, "simul et iucunda et idonea dicere vitae" (Horace 1942, 478) [to utter words at once pleasing and helpful to life (Horace 1942, 479)]. Allusions to this ideal are ubiquitous in novels of the seventeenth and early eighteenth centuries and figure prominently also in Lesage's *magnum opus*, *Histoire de Gil Blas de Santillane*. But what is crucial here is that the dichotomy of pleasure and usefulness corresponds directly to the dichotomy of visibility and invisibility, description and narration. The mere seeing of the presently visible scenes is delightful; but the contemplation of these scenes remains useless unless one hears the story of the events leading up to these scenes. The devil renders the "amusement frivole" [frivolous amusement] of pure seeing useful by explaining what the people are doing and why they are doing it. And the latter task involves also an explanation of what the presently visible people previously did. Explaining the visual in *Le Diable boiteux* means, in other words, telling the story (*histoire*) – long or short – of that which happened before. And these stories themselves remain invisible. Useful narration, in *Le Liable boiteux*, is almost completely divorced from the frivolous amusement of the visible.

Again and again, the student's pleas to explain a particular scene result in the devil's telling the story of what happened before. The beginning of the first longer novella, which is introduced by the student's question about a particular tableau, illustrates this structure:

> Expliquez-moi, de grâce, interrompit Léandro Perez, un autre *tableau* qui se présente à mes yeux. Tout le monde est encore sur pied dans cette grande maison à gauche. D'où vient que les uns rient à gorge déployée, et que les autres dansent ? On y célèbre quelque fête apparemment ? Ce sont des noces, dit le boiteux. Tous les domestiques sont dans la joie. Il n'y a pas trois jours que dans ce même hôtel on était dans une extrême affliction. C'est une *histoire* qu'il me prend envie de vous raconter : elle est un peu longue, à la vérité ; mais j'espère qu'elle ne vous ennuiera point. En même temps il la commença de cette sorte. (Le Sage 1984, 51–52, my emphasis)

> Explain to me, if you will, interrupted Léandro Perez, another *tableau* that presents itself to my eyes. They are all still up on their feet in this big house to the left. Why is it that some are laughing with their mouths wide open, and that others are dancing? One is celebrating here apparently? It is a wedding, said the devil. All the servants are happy. It hasn't been quite three days that there was great affliction in this house. It is a *story* that I am in the mood to tell you. Admittedly, the story is a bit long; but I hope that it will not bore you. At the same moment, he started in the following way. (My emphasis)

In Lesage's novel, the description of the visible tableaux always needs to be supplemented by the narration of the backstory. While description is concerned with the present visuality, narration delivers the fundamentally invisible story that reveals the truth behind the present scene.

Although description and narration seem thus entirely opposed to one another, they are actually united by the same logic of discovery. The visual scenes appear as a consequence of the uncovering of the roofs – a laying bare that impresses the visual images upon the student (and the reader). The devil's narratives add to this first instance of discovery a second discovery in which the truth behind the images is revealed: many of the novel's scenes are concerned with deceit and false appearances. Take, for instance, the following scene, which appears early on in the novel and in which the student believes to see "une grande et jeune fille faite à peindre" (Le Sage 1984, 42) [a tall and young girl fit for painting]. The devil's answer serves to reveal the truth behind the misleading appearance:

> Hé bien, reprit le boiteux, cette jeune beauté qui vous frappe est sœur ainée de ce galant qui va se coucher. [...] Sa taille, que vous admirez, est une machine qui a épuisé les

mécaniques. Sa gorge et ses hanches sont artificielles ; et il n'y a pas longtemps qu'étant allée au sermon, elle laissa tomber ses fesses dans l'auditoire. (Le Sage 1984, 42)

> Oh well, replied the staggering devil, this young beauty that strikes you is the older sister of that suitor who is now going to bed [an equally vain old man, who also tries to appear young; described briefly before]. Her waist, which you admire, is a machine whose mechanics are exhausted. Her breast and her hip are artificial, and not too long ago, when she went to church, her buttocks fell on the floor amid the audience.

In a scene such as this one, the narrative is, to be sure, only minimal. And still it seems significant that the devil does more than simply correct the student's description. He does not simply say that the woman is not as young as she looks, thus substituting the student's wrong description with a more accurate one. Over and above such a simple correction, the devil's statement culminates in a short narrative of a past event in which the truth is revealed: in this past story, the woman's artificial buttocks fall to the floor and thus literally expose her act of deception.

Making images

However, before the narration of the stories can begin to explain what is presently visible, the visual has to be organized so that there appear clearly demarcated and describable tableaux to which the devil's stories can refer. The existence of such clearly framed individual tableaux is by no means given from the beginning. When Don Cléofas distractedly looks over the newly uncovered houses at the outset of the novel, there are no tableaux yet. There is, instead, only a "confusion d'objets" [confusion of objects]: the pure sensation of data that potentially could be described and observed.

The first necessary step toward the narration of the visual is thus the creation of tableaux. There needs to be a technique for the construction of such tableaux. Devil and student need to carve out of the confusion of sensory data the objects that they set out to describe. Only subsequently can they begin to describe what they see and to embellish their descriptions by narration. This is a problem that I have paid little attention to so far. When I discussed the procedure of observation in the introduction to this book, I largely focused on cases in which an image suddenly appears and strikes the attention of a homodiegetic observer (and, by extension, that of the reader). But the situation in *Le Diable boiteux* is different from, say, that of the observer at the beginning *Madame Bovary*. Instead of reading a direct account of what is seen, *Le Diable boiteux* presents us with a dialogic situation in which vision is mediated by the conversations between devil and student. Rather than simply having images appear, devil and student have to agree together on an image that they can subsequently comment on. More precisely, they have to visualize for each other the image that will become the object of the subsequent narration. In some way, this is one of the ultimate jokes of *Le Diable boiteux:* that this novel, which seems to place so much value on the visual discovery of the world, actually presents visuality only indirectly, as a product of the conversation between the devil and student. The novel, in other words, starts again and again, from the visual images of description, but these very images are presented within a framework of *storytelling* (a construct that, as we shall see, plays an even more dominant role in *Les Nuits de Paris*).

In having the images appear as a product of the conversations between devil and student, Lesage's novel allows us to study how it is that literature produces effects of visualization. *Le Diable boiteux*, in other words, is a novel that sets out from individual visualized images, and it offers us, in the dialogues between devil and student, also a poetological guidebook for how to produce such images through language.

The way in which the production of intersubjective visual objects is achieved in *Le Diable boiteux* is, first and foremost, through a compartmentalization of the field of vision. The devil intervenes in the student's confused contemplation of the landscape of unroofed houses to define clearly demarcated spaces that can be described. The devil defines such spaces first and foremost by simply *pointing* to this or that room among the multitude of rooms that are presently visible. Mundane as the devil's deictic gestures appear, they are central for the procedures of description and narration in the novel. The devil *dictates* the course of the novel by constructing describable objects through his *deictic* gestures.

Figure 4: The devil pointing to individual rooms in the city; detail of an engraving by Dubercelle (see Fig. 2). Source gallica.bnf.fr / BnF.

Guided by the devil's deictic gestures, devil and student advance in their spectatorship house by house, room by room, and person by person. Lesage rarely misses a chance to introduce a scene without mentioning the deictic gestures and imperatives through which the devil achieves this compartmentalization of the field of vision

- Observons d'abord dans cette maison à ma droite [...]. (Le Sage 1984, 41) [Let us begin our observations in this house on our right (...).]
- [P]renez garde [...] à ce qui se passe dans une petite salle de la même maison. (Le Sage 1984, 41) [Pay attention (...) to what happens in a little room of this house.]
- Jetez les yeux sur cet hôtel magnifique [...]. (Le Sage 1984, 43) [Watch that great townhouse (...).]
- Regardez un peu au-delà [...] et considérez dans une salle basse [...]. (Le Sage 1984, 43) [Regard a bit over there [...] and consider this room at the bottom (...).]
- Portez la vue au-delà, sur la droite, et tâchez de découvrir dans un grenier [...]. (Le Sage 1984, 45) [Watch over there, on the right, and try to discover in the attic (...).]

These deictic gestures contain an imperative to see, and as such Lesage relies on a wisdom known since antiquity – that it is through references to the act of seeing that one produces effects of visuality (Innocenti 1994, 374–375). But aside from this technique of producing visual images by the inclusion of an imperative to see ('look here and here and here'), another important feature of the novel's deictic gestures is the division of the space into distinct rooms. In Lesage's novel, the city of Madrid appears as a huge dollhouse in which we can calmly pass from room to room. And each room functions as a separate stage setting to which one can point and in which one can place a narrative.

The novel includes several strategies to preserve the rooms as closed-off units and to support the devil's organization of space through his deictic gestures. First of all, *Le Diable boiteux* is set at nighttime, when there is little interaction out on the streets. Moreover, the novel includes sections of considerable length in a prison and in a mad-house – places in which there is by definition no interaction between the individual cells. The compartmentalization of space through deictic gestures, which is necessary for the creation of describable objects, is thus prefigured in the space in which the spectatorship of devil and student takes place.[6]

---

6 Volker Klotz analyzes the novel's multiple ways of restricting movement between clearly demarcated rooms in his 1969 study *Die erzählte Stadt* [The Narrated City]. While Klotz praises *Le Diable boiteux* as the first novel to bring the city onto the stage of the novel, he mainly stresses that Lesage lacked a sense of how to describe the entanglement and steady movement of the

The reinforcement of the compartmentalization of the city-space through the devil's deictic gestures is the decisive technique for the construction of describable tableaux in *Le Diable boiteux*. Another recurring procedure that similarly serves the production of describable objects is one of 'disentanglement.' The word that is used at various points in the novel is "démêler" [to sort out, separate, unknot]. Like the deictic gestures, the main function of these imperatives to disentangle seems to be their implicit cue to visualization – a form of visualization, more precisely, in which a figure takes shape against a blurry background:

- De mon côté, dit Asmodée, je considère trois ombres remarquables, que je démêle dans la foule. (Le Sage 1984, 181) [To my side, said Asmodée, I see three remarkable shadows that I have singled out (*démêle*) from the crowd.]
- J'entends ronfler autours de nous, dit Léandro Perez, et je crois que c'est ce gros home que je démêle dans un petit corps de logis [...] (Le Sage 1984, 258) [I hear snoring around us, said Léandro Perez, and I believe that it is this fat man which I can make out [*démêle*] in this small building (...).]
- Je démêle dans cette salle un poète qui n'y devrait pas être. (Le Sage 1984, 281) [I am spotting (*démêle*) in this room a poet who should not be there.]

Demarcating a space of vision and singling out individuals against a blurred background – this is how describable tableaux are created in *Le Diable boiteux*. Lesage's novel does not comment on these procedures in any great detail, but it does show them in some minimal form persistently in preparation of almost all of the dozens and dozens of scenes. And through these many repetitions, the novel presents us with something like an implicit poetology of visualization, telling us how to produce images through language.

The description that follows the identification of the tableaux remains minimal in *Le Diable boiteux*. There are no such detailed accounts of physiognomies, clothes, or furniture as we find them in nineteenth-century novels by Dickens, Flaubert, or Stifter. In fact, in his essay "Erzählen oder Beschreiben?" Georg Lukács explicitly mentions Lesage as a representative of eighteenth-century literary style, in which description plays only a minor role:

---

modern city's active crowds. Klotz speaks of "zu Lesages Zeit noch kaum darstellbaren vielfältigen Verästelungen der Stadtbevölkerung" (Klotz 1969, 27) [complex entanglements of the urban population, which were not possible to represent in Lesage's time] and explains Lesage's attempts to represent the city as "Gewalttakt[e], wo spätere Romanciers ungleich geschmeidiger vorgehen können" (Klotz 1969, 46) [measures of blunt force at points where later novelists can proceed incomparably more smoothly]. For similar arguments see, more recently, Reiffers 2013, 255–256.

Der Roman des achtzehnten Jahrhunderts (Le Sage, Voltaire u. a.) hat die Beschreibung kaum gekannt; sie spielte in ihm eine verschwindende, eine mehr als untergeordnete Rolle. Erst mit der Romantik ändert sich die Situation. Balzac hebt hervor, daß die von ihm vertretene literarische Richtung, als deren Gründer er Walter Scott ansieht, der Beschreibung eine größere Bedeutung zuweist. (Lukács 1955, 109)

In the novel of the eighteenth century (Le Sage, Voltaire, etc.) there had scarcely been any description, or at most it had played a minimal, scarcely even a subordinate role. Only with romanticism did the situation change. Balzac pointed out that the literary direction he followed, of which he considered Walter Scott the founder, assigned great importance to description. (Lukács 1970, 117)

But Lukács's extreme opposition between Lesage and nineteenth-century writers is not entirely accurate. As Marguerite Iknayan has shown, in the nineteenth century, Lesage, although he includes relatively little description compared to later writers, was actually perceived as a bridge to the contemporary writers' emphasis on description:

Lesage still belonged chiefly to the classic school but was at the same time a writer of transition tending toward the representation of more specific exterior detail. As such he helped prepare the growing taste for individual and local reality which Walter Scott was doing so much to foster in the 1820's. (Iknayan 1958, 374–375)

Lesage, in other words, was instrumental in introducing longer descriptive passages in literary narratives, even if there seems to be, from a later perspective, still only very little description in his works. In *Le Diable boiteux*, the description of the presently visible tableaux merely serves to furnish the scenes with a minimal inventory of objects and persons that appear in the stories. The devil subsequently supplements these descriptions with narratives that connect and explain the described persons and objects. The descriptions alone neither fully portray the characters – as the descriptions of many nineteenth-century realists do – nor do the descriptions directly blend over into the narrative, as is the case in the procedures of observation that I study in this book.

Observation in *Le Diable boiteux*

In saying that Lesage's novel does not contain observation proper, I do not mean to suggest that the term observation is absent from the novel. The French verb "observer" appears several times in the novel. But it is used more or less indiscriminately with other terms of visual engagement, including "contempler," "considerer," "regarder," and "examiner." All these words refer to the devil's and student's activity of watching the tableaux. Like these other verbs, "observer" in *Le Diable boiteux* means to look at a tableau that has been singled out of the mass of available visual data – as in "Observons d'abord dans cette maison à ma droite ce vieillard qui compte de l'or et de l'argent" (Le Sage 1984, 41). [Let us first observe in this house on my right this old man who is counting gold and silver.]

This usage of the verb "observer" in *Le Diable boiteux* is largely in line with the general usage of "observation" in early eighteenth-century French. At this time, "observer" evokes first of all the physician's examination, or "observation" of the patient lying in bed. Along these lines, Lesage employs in his novel *Histoire de Gil Blas de Santillane* repeatedly the phrase "observer le malade" (Le Sage 1973, 130, 145, 149) [to observe the sick person]. Even in the second half of the eighteenth century, in Diderot and d'Alembert's *Encyclopédie*, the entry "observation" is still largely concerned with the practice of observation in medical treatment. Lesage's devil and student in *Le Diable boiteux* 'observe' the tableaux, as a doctor 'observes' a patient who lies motionlessly in bed.

In contrast to some of the later texts that I analyze in this book, *Le Diable boiteux* includes only very few scenes of what I define as observation. The visual in Lesage's novel appears, for the most part, only in the form of static tableaux of individual moments. There are at least two important cases, however, in which we are presented with what I call an observation – a scene in which a person (or object) is described and subsequently shown in its dynamic development over time. The first instance appears already on the very first page of novel. Indeed, we are introduced to the novel's protagonist, Don Cléofas, through an observation – in a way that might recall (in the context of this present study) the way in which Charles Bovary is introduced at the beginning of *Madame Bovary*:

Une nuit du mois d'octobre couvrait d'épaisses ténèbres la célèbre ville de Madrid : déjà le peuple, retiré chez lui, laissait les rues libres aux amants qui voulaient chanter leurs peines ou leurs plaisirs sous les balcons de leurs maîtresses : déjà le son des guitares causait de l'inquiétude aux pères, et alarmait les maris jaloux : enfin il était près de minuit, lorsque don Cléofas Léandro Perez Zambullo, écolier d'Alcala, sortit brusquement par une lucarne d'une maison où le fils indiscret de la déesse de Cythère l'avait faire entrer. Il tâchait de conserver sa vie et son honneur, en s'efforçant d'échapper à trois ou quatre spadassins qui le suivaient de près pour le tuer, ou pour lui faire épouser par force une dame avec laquelle ils venaient de le surprendre. (Le Sage 1984, 29)

One October night covered the famous city of Madrid in thick darkness: the people, already retired in their homes, left the streets to the lovers who wanted to sing of their pain or their pleasure under the balcons of their mistresses; the sound of guitars was already disquieting the fathers and alarming the jealous husbands; finally, close to midnight, Don Cléofas Léandro Perez Zambullo, student at the Alcala, suddenly got out from an attic window of one of the houses that the indiscrete son of the goddess Cythera [i. e. Cupid, the son of

Venus] had allowed him to enter. He tried to conserve his life and honor, struggling to escape from two or three hired murderers, who followed him closely to kill him or to force him to marry a lady in whose company they had just surprised him.

Lesage opens his novel with the theatrical stage setting of the comedy of manners. Night falls over the city, the people are withdrawn in their houses, and gradually, one after the other, the classical prototypes of the comedy of manners appear: happy and unhappy lovers, watchful fathers, and jealous husbands. It is against the backdrop of this tableau that "suddenly" (*brusquement*) the protagonist appears – rushing onto the stage from one of the city's attics. To be sure, Lesage does not offer a clear description as to how this protagonist looks (and in this sense, there is a marked contrast to the observation at the beginning of *Madame Bovary*). The only 'description' we are offered consists in the lengthy name (don Cléofas Léandro Perez Zambullo) and in the specification that he is a student at the prestigious university of Alcala near Madrid – but maybe this is already enough to trigger an image. Following this minimal image is a slightly longer sequence in which we see the student run over the roofs of the houses and finally enter another attic, in which he encounters the devil Asmodée.

This opening sequence and the observation that it contains would not normally be very remarkable, were it not for the fact that this observation stands in sharp contrast to almost the entire rest of this novel, in which the visual descriptions of the scenes under the roofs of the city are categorically separated from the narratives (which are provided by the devil) – we hardly ever again see the transition from the sudden appearance of a momentary image to the extended watching of how this image develops over time. The only other important exception to this pattern occurs exactly in the middle of the novel. At this point, the devil intervenes for the first and only time directly in one of the scenes that he and the student witness.[7] In this episode, Asmodée and Don Cléofas fly to the burning house of Don Pedro, a wealthy citizen of Madrid. At the moment the two spectators arrive, Don Pedro's house is being consumed by flames and his daughter Séraphine is captured within it. Don Pedro cries in vain for people to rescue his daughter from the burning house. No one is willing to risk his life to save Séraphine, beautiful and rich as she may be. Don Cléofas, however, moved by the misery of the young woman, implores the devil to rescue the girl. After some mocking and reluctance, the devil agrees to do it. In the guise of Don Cléofas, he makes his way into the burning building and carries

---

7 Once before that, the devil incites a fight between a group of assassins (Le Sage 1984, 100) who had attempted to kill the student Don Cléofas at the outset of the novel (Le Sage 1984, 29). However, in this case the devil does not physically appear on the stage of the watched scene.

Donna Séraphine out of danger. Meanwhile, Don Cléofas observes the devil's intervention from a safe distance. Séraphine's father is deeply grateful for what he takes to be Don Cléofas's courageous deed. In recompense, he later, at the end of the novel, grants Don Cléofas the hand of his daughter.

The rescue scene in the middle of the novel is structurally of great importance to *Le Diable boiteux* because it ties the frame narrative about the spectators to the myriad of scenes that the two spectators witness. It is in the rescue scene, as the German critic Karl Riha correctly remarks, that the collection of stories is transformed into a novel proper (Riha 1992, 21). And it is in this structurally important scene that the novel introduces, moreover, a scene of literary observation. Before the devil actually assumes the figure of Don Cléofas and rushes into the burning house, he instructs the student to carefully "observe" his every action: "[R]egardez de quelle façon je vais m'y prendre. *Observez* d'ici toutes mes opérations." (Le Sage 1984, 171, my emphasis) [Watch how I will behave; *observe* from here all my actions. (My emphasis)] In contrast to all previous scenes, in which observation simply means looking at a particular tableau, observation includes in this case a sustained watching of a series of actions unfolding over time. And whereas in all previous scenes the devil needed to supplement the tableau with an explanatory narrative, in this case observation and narration collapse into one. The student's observation of the devil's deed and the narration of the rescue scene coincide.

Looking at the very opening of this novel and the rescue scene in the middle of the novel, one cannot help but wonder why Lesage had no qualms about including observations in the frame narrative, while at the same time avoiding them in the account of what devil and student see under the roofs of the houses (i.e. the framed stories). One possible way to explain this curious phenomenon may be to say that there are two different regimes of visual perception governing the novel, and that it is only within the moral regime of uncovering (or revelation) in the framed stories that observation is impossible. The regime of uncovering, which limits the visual to single images, is in place when devil and student look at the scenes under the roofs of the houses. But it is not in place in the frame narrative, when we look with the narrator at devil and student themselves. In making this difference between frame narrative and the framed stories, Lesage does much to strengthen one of the main underlying assumptions of the present study, namely that literary observations depend on very specific formal and cultural preconditions. As we can learn here from Lesage's novel, at least in the moral regime of uncovering, observation remains problematic, at best. Over the course of the following chapters, we will encounter a range of other epistemic frameworks that similarly either support or inhibit literary observations.

The legacy of *Le Diable boiteux*

Le Diable boiteux assumes in many ways an ambivalent position vis-à-vis the epistemic potential of the visual. On the one hand, the novel prominently introduces the visual as the beginning of all storytelling, and it offers basic insights into the production of visible objects, or tableaux, out of the initial confusion of visual data. On the other hand, visuality remains in the majority of scenes associated only with describable tableaux, not with the narratives, which are told by the omniscient devil. As a consequence, literary observations remain the exception in *Le Diable boiteux*. The visual is both enthroned and limited by the same logic of a revelation of human shortcomings. Lesage's novel, more precisely, is governed by a literary and moral concept of a corrupt humanity that hides its vices, and this concept inhibits the reliance of the visual as much as it promotes it.

In the history of reception of *Le Diable boiteux* in the eighteenth and nineteenth centuries, the novel's ambivalent assessment of the visual is essentially resolved. The devil Asmodée's lifting of the roofs simply becomes one important reference point for a fiction of absolute visibility and the possibility to narrate stories through the engagement with the visual. The limitations that characterize the visual in *Le Diable boiteux* are removed. E.T.A. Hoffmann's 1815 story "Der Dey von Elba in Paris. Sendschreiben des Türmers in der Hauptstadt an seinen Vetter Andres" [The Dey of Elba: Missive of the Watchman in the Capital to his Cousin Andres] is a case in point. In "Der Dey von Elba in Paris," Lesage's devil Asmodée reappears to help a Berlin watchman look into the houses of the city. But instead of providing merely static images in need of explanatory narratives, the devil allows the Türmer to observe entire sequences and even to overhear conversations. Moreover, this story hardly raises the question, central to *Le Diable boiteux*, of whether the disclosed scenes inside the houses are deceptive. Lesage's devil, in other words, is reduced to a cipher of unlimited visibility in Hoffmann's story. Something similar still holds true for Arthur Conan Doyle's early Sherlock Holmes story "A Case of Identity" (1891), which I discuss in more detail in the final chapter. In a conversation with Watson, Sherlock Holmes imagines the two of them flying over London, lifting the roofs of the houses, and watching the events inside as they unfold over generations (Doyle 1953, 190 – 191). The fact that precisely such an unfolding of events is impossible to observe in Lesage's novel is ignored in Doyle's story. Rather than being received as the fundamentally ambiguous novel of the visual that it really is, *Le Diable boiteux* became, instead, a cipher for a form of storytelling that is firmly grounded in an encounter of the world as it visually appears to us. In some sense, we might even say, *Le Diable boiteux* becomes a template for a storytelling based on observation – in the technical sense of the term that I rely on in this book: as a form of writing that extends the visuality of description into the narrative.

It is worth defining the legacy *Le Diable boiteux* in a little more historical detail as it helps us demarcate the age in which procedures of observation are an important literary concern. Once the interest in *Le Diable boiteux* abates, I argue, the attempt to mimic in literature a visual encounter with the world (a seeing of the world, in other words) also loses importance. Strikingly, the reception history of *Le Diable boiteux* has a rather precise endpoint. Published at the dawn of the eighteenth century and an immediate success in the bookstores, *Le Diable boiteux* was reprinted every couple of years over the following decades. Illustrated editions of *Le Diable boiteux* appeared throughout the eighteenth century and particularly also in the nineteenth century. Between 1790 and 1890, almost every year saw a new French edition of *Le Diable boiteux* – until, suddenly, the series of new editions stopped at the fin-de-siècle (see Fig. 5). The history of English translations of *Le Diable boiteux* shows a similar development – with the only difference that we see here no clear difference between the eighteenth and nineteenth centuries. We have at least eighteen English editions between 1708 and 1793, and at least another eighteen between 1800 and 1900. After the turn to the twentieth century, however, there is no new English edition until 1972.[8]

The history of pastiches and adaptations of *Le Diable boiteux*, as well as of references to the novel in later fiction, presents us, once more, with a comparable development (see Saint-Amour 2011, Meglin 1994, Arac 1979). There was an immediate literary response to the novel at the time of its first publication and a thriving interest over the nineteenth century. In German literature, we find references to *Le Diable boiteux* in works by Wieland,[9] Goethe (in at least three different texts),[10] and E.T.A. Hoffmann.[11] In France, Balzac signed several of his texts

---

8 These numbers are based on the data available on worldcat.org, which remains in some cases ambiguous (for instance, when the publisher is not listed and it therefore cannot be determined whether an entry reflects a distinct edition) and which may be incomplete. In most cases, a new edition is not at the same time a new translation. Curiously, the history of German translations of *Le Diable boiteux* shows us a surprisingly different dynamic. After relatively few editions in the eighteenth century, the number of German editions picks up in the nineteenth century, and we continue to see several new German editions in almost every decade of the twentieth century – with the exception only of the 1930s and 1940s.

9 Wieland briefly turns to *Le Diable boiteux* in his 1771 comic verse epic *Der neue Amadis* [The New Amadis] (Wieland 1839, vol. 15, 323).

10 There is a short reference to Lesage's novel in Goethe's autobiographical writings *Dichtung und Wahrheit* [Poetry and Truth] (Goethe 1986, 24) and *Campagne in Frankreich 1792* [Campaign

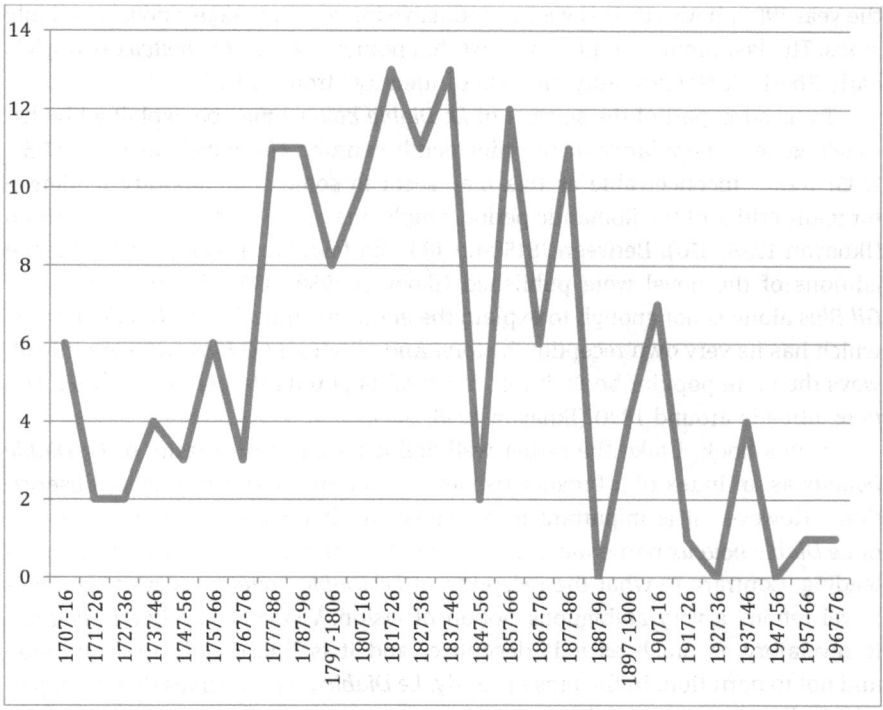

Figure 5: Number of new French editions of *Le Diable boiteux* per decade from 1707 to 1976. Based on an analysis of the entries on worldcat.org.

as "Le Diable à Paris" [the devil in Paris] (Saint-Amour 2011, 230), and for British and American realist writers such as Thomas Carlyle, Charles Dickens, Henry Wadsworth Longfellow, and Nathaniel Hawthorne, Lesage's novel was an important point of reference (Saint-Amour 2011, 231). Examples of pastiches by lesser-known writers include *The Devil Upon Crutches in England: Or, Night-Scenes in London* (1756) and *Asmodeus in New York* (1868). In fact, by the end of the nineteenth century there was a "diable boiteux" for many of the larger cities of the Western hemisphere, including *Asmodeus: Or, The Devil in London* (1808), *Der hinkende Teufel zu Berlin* [The Staggering Devil in Berlin] (1827), and *Der hinkende Teufel in Hamburg* [The Staggering Devil in Hamburg] (1840). Around

---

in France in the Year 1792] (Goethe 1994, 534), as well in his play *Die Vögel* [The Birds] (Goethe 1988, 229).

11 See my brief discussion of Hoffmann's story "Der Dey von Elba in Paris" (1815) above.

the year 1900, however, this vastly productive interest in Lesage's novel suddenly stops. The last prominent fictional text that points to *Le Diable boiteux* is Doyle's early Sherlock Holmes story "A Case of Identity" from 1891.[12]

To be sure, part of the success of *Le Diable boiteux* may be explained by Lesage's general popularity in the nineteenth century. His novel *Gil Blas* (1715–1735) was – inconceivable as this may seem to some contemporary readers – for many critics of the Romantic period simply one of the best novels ever written (Iknayan 1958, 370). Between 1815 and 1830 no fewer than twenty-eight French editions of the novel were published (Iknayan 1958, 370). But the success of *Gil Blas* alone is not enough to explain the enduring appeal of *Le Diable boiteux*, which has its very own reception history. And although *Gil Blas* was certainly always the more popular book, it lost much of its popularity before *Le Diable boiteux*, already around 1830 (Iknayan 1958, 377).

In this book, I take the rather well-defined arch of popularity of *Le Diable boiteux* as an index of a broader rise and fall of an interest in literary observations. However, it is important to remember in this context that the reception of *Le Diable boiteux* perpetuates an image of Lesage's novel that is strikingly misleading. Contrary to what the reception of *Le Diable boiteux* suggests, Lesage's novel retains a very ambiguous notion of vision. Vision, in *Le Diable boiteux*, is revelatory as much as it is deceptive, and it is linked only to description and not to narration. In the present study, *Le Diable boiteux* serves therefore, paradoxically one might say, both as a historical starting point of an interest in literary observation and as one specific case study of a literary and cultural constellation in which observations remain impossible. Lesage's novel shows us, by way of negation, some of the preconditions on which the seemingly so simple procedure of observation relies. In the next chapter, I will look at a text in which the limitations on the visual that characterize *Le Diable boiteux* are lifted and in which the procedure of observation is full on display.

---

12 Paul K. Saint-Amour claims to find an instance of a "straightforwardly Asmodean Vista" (Saint-Amour 2011, 233) in James Joyce's *Ulysses* (1922) and even "an explicit engagement [...] with the figure of the *diable boiteux*" (Saint-Amour 2011, 243) in *Finnegans Wake* (1939). However, the connection to Lesage in *Ulysses* seems rather tenuous, and the brief reference in *Finnegans Wake* is, strikingly, no longer to Lesage directly but, instead, to Doyle's Sherlock Holmes story "A Case of Identity." Looking beyond literature, one might be tempted to place the opening sequences of Wim Wenders's film *Der Himmel über Berlin* [translated as Wings of Desire] (1987) in this tradition.

Chapter 3: Observation (*Les Nuits de Paris*)

In the previous chapter, I discussed the turn to the visible world as the basis of all storytelling in Alain-René Lesage's novel *Le Diable boiteux*. In Lesage's novel, the devil Asmodée takes a student on a flying tour over the city of Madrid and lifts the roofs of the houses, so that the student enjoys a seemingly unobstructed view of the interior scenes. But understanding these scenes requires that the devil supplies a narrative of the backstory and, importantly, these backstories themselves are not visible.

If the devil of Lesage's *Le Diable boiteux* has earned a place in literary history for his gesture of grounding all stories in the visible world, Rétif's nocturnal spectator in the novel *Les Nuits de Paris, ou le Spectateur-nocturne* (1788), which I analyze in this chapter, equally deserves a place for actually observing the reality to which Lesage's devil only superficially points.[1] The nocturnal spectator's observations on his nightly walks through Paris stage full stories as residing within a world that is disclosed in and by virtue of its visibility. Moreover, the technique of observation, which in most works – even of literary realism – is just a tool that is used more or less sparingly at individual (albeit important) points, turns in *Les Nuits de Paris* into a central concern of the entire novel. Consider here once more, for the sake of contrast, the opening of *Madame Bovary*. In this opening scene, we certainly have a prominently placed observation that reveals much about the concerns of this novel (especially about the struggle to provide a literary portrait of a character like Charles Bovary). But it is not the case that the rest of Flaubert's novel relies in a similarly strong way on such instances of observation. *Les Nuits de Paris*, however, consists of little else than a constant repetition and variation of such scenes of observation.

To be perfectly clear, my interest here is in the narrative techniques that simulate the visuality of reality and that perform, through the nocturnal spectator in the text and for the reader, processes of perception that move from the seeing of images to the seeing of sequences. I am not much interested in the question of whether or not the content of the nocturnal spectator's observations is actually based on the recording of any true extra-textual reality. Nonetheless, it may not

---

1 Rétif's *Les Nuits de Paris* still awaits major critical discussion. In recent years, however, *Les Nuits de Paris* – as well as Rétif's colossal oeuvre, which fills "an estimated 57,000 pages and 187 volumes" (Wyngaard 2013, 2), in general – has experienced a certain rise in interest. For an overview of the reception of Rétif's works, see Wyngaard 2006 and Wyngaard 2013, 1–11. The status of the scholarship on *Les Nuits de Paris* is summarized in Mall; for more recent discussions of *Les Nuits de Paris*, see esp. Barr 2012, Goulemot 1989, Klein 1994.

https://doi.org/10.1515/9783110594348-091

be entirely amiss to note that, in contrast to Lesage's novel, such a documentary aspect is for parts of Rétif's novel quite possible and even likely. At least in the case of this novel, the narratological technique of observation seems to be expanded alongside the actual observation of the world.[2] Be this as it may, what is most important to me is the new narratology of observation that we can find in *Les Nuits de Paris*. In contrast to Lesage's novel, where the presently visible and describable tableaux under the roofs of the houses form, for the most part, only the starting point for the devil's narration of the events that occurred prior to the visible tableaux, Rétif's text continues the visual encounter of the tableaux in the sustained recording of sequences that also allows the construction of the narrative as a *visual* reality. Importantly, the novel shows that the constitution of this complete visual reality relies on a literary procedure that combines description with narration. Moreover, *Les Nuits de Paris* provides us also with a theory of its own narrative procedure. The novel reflects in its many stories on its own structure of an ideal observational procedure, which proceeds from the initial openness to be struck by a *describable* image to the subsequent focused watching of a *narrative* sequence of events.

After reconstructing Rétif's ideal observational procedure in this chapter, I will show in the next chapter through readings of texts by Goethe, Büchner, and Poe that many writers of the late eighteenth and early nineteenth centuries worked to problematize precisely this ideal procedure. This is not to say that all these writers were directly responding to Retif's novel (most likely they were not), but that the general question of how to construct stories – and reality – as observations was 'in the air' at this point. As my readings of Goethe, Büchner, and Poe show, the most difficult step in this procedure lies in the transition from the unfocused gaze of the spectator, which provides the images that are the starting point of the observations, to the sustained and focused gaze, which produces a narratable sequence. In Goethe's novel *Die Leiden des jungen Werthers*, the problem consists, schematically speaking, in the fact that Werther repeatedly focuses on a single object or tableau, but fails to account for the change that the tableau undergoes as time passes. In Georg Büchner's novella *Lenz*, by contrast, the problem consists in the integration of the changing images into a coherent narrative sequence. In Edgar Allan Poe's short story "The Man of the Crowd," finally, the urban spectator fails even to focus on any one the people that he sees – which alone would allow for the observation of an extended sequence. Goethe's, Büchner's, and Poe's cases of observational failure will help us to appreciate the

---

2 William Edmiston offers a brief summary of the debate over the fictionality of *Les Nuits de Paris* (Edmiston 1994, 47, 62).

complexity and the rich cultural preconditions of a procedure whose basic and seemingly simple mechanism is showcased in *Les Nuits de Paris*.

Before looking at the complications in Goethe's, Büchner's and Poe's texts, however, we should be clear about the ideal structure of observation on which, implicitly, these texts comment. Rétif's novel *Les Nuits de Paris* stages this structure again and again during the 381 nights that the nocturnal spectator spends wandering through the city. The novel carefully distinguishes the steps of the process in which storytelling emerges out of the observation of the visible world. In analyzing this process, I will focus on something like the x-ray image of the narratological and epistemological structure on which Rétif's novel relies. As a consequence, I will pay comparatively little attention to the rich content of the 3000-page novel, and I will also pass over the – in another context – interesting question of how Rétif's nocturnal spectator, who traverses the streets of late-eighteenth-century Paris, relates to the literary and social figure of the *flâneur*, who emerges in the first decades of the following century. Suffice it to say that there certainly does seem to exist a connection between the nocturnal spectator and the flâneur, though not perhaps a connection to the Benjaminian and Baudelarian *homme des foules*, who "convey[s] the shock effect of modernization" (Lauster 2007, 148), but instead to a simpler type who walks and observes the city.[3]

In the first step of the observational process (analyzed in the first section of this chapter), the nocturnal spectator becomes drawn to a curious tableau. Walking through Paris, the nocturnal spectator initially does not focus on anything in particular. Instead, he is open to being struck by anything that might be worth watching. This still unfocused wandering is what defines *spectatorship*. The unfocused spectator is fundamentally limited to the seeing of describable tableaux, because the seeing of sequences would demand that the spectator has already started devoting his attention to the sustained watching of one single scene. The fact that the novel's protagonist is called nocturnal *spectator* stresses the importance of this preparatory stage of observation.

In the second step (analyzed in the second section), however, the initial encounter with a curious momentary tableau is transformed into a sustained and focused recording. While it is the continued and focused recording that discloses the sequences of events that constitute the novel's stories, the stories would not exist without the previous seeing of an interesting tableau. Only a curious ta-

---

3 For a critique of Benjamin's concept of the flâneur and diverging, historically earlier concepts of the flâneur, see Lauster 2007, 147 and Burton 1994, 4. For a discussion of the relation between Rétif's novels and the literature of *flânerie* see Turcot 2006 and Turcot 2010.

bleau can attract the attention of the unfocused gaze of the spectator. At the same time, the seeing of the tableaux alone is insufficient for the telling of the stories. The initially described tableaux lend themselves to the telling of a story only once the nocturnal spectator begins recording over time what he initially and momentarily perceived. Observation consists in the combination of these two distinct visual processes – the initial openness to be struck by an image and the subsequent focused watching of a sequence of events. In Rétif's novel, these two distinct visual processes are represented through description and narration respectively. The novel thus shows us that the construction of a visual reality relies on the combination of description and narration, and the novel helps us refine traditional accounts of literary realism in which visualization is understood as a feature of description alone.

However, as I discuss in the third section of this chapter, even this combination of initially perceived tableau and continuously watched sequence of events in the observational process falls short of being a story proper. As Rétif's novel stresses again and again, the observed sequences rely on an act of storytelling that, while feeding off observation, is distinct from observation itself. The nocturnal spectator, who observes strange occurrences in the streets of Paris, also explicitly acts as a storyteller to an intradiegetic and extradiegetic audience to whom he relates his observations. As I have already discussed in some detail in the introduction, I use the term 'storytelling' for the act of relating a story, in distinction to the technique of 'narration' which is, in my definition, part of the observational processs. 'Storytelling' encompasses both description and narration. Only to the extent that there is an observed reality can there be stories in *Les Nuits de Paris*, but conversely, the observed reality becomes thematic only in a framework of storytelling.

With this third step, Rétif offers us an important elaboration of the relation between the empirical practice of observation and the form of literary representation. On the one hand, the main impetus of Rétif's novel seems to be to ground the stories in the visual – to legitimize literature as observation. On the other hand, however, Rétif also asks whether the novel really only 'translates' a visual process of observation into its appropriate form of literary representation (description plus narration), or whether storytelling has not itself, in some form, priority. The question that is raised, in other words, is that of whether observation is not from the outset structured in the form of storytelling. And this preexisting framework of storytelling also complicates the idea of 'openness' as a central feature of the observational process. Observation in literature moves within the confines of the closure of storytelling.

Spectatorship

Let us first look at the preparatory stage of observation, in which the observer does not yet focus on anything. Historians of science sometimes overlook this stage. Ralf Klausnitzer, for instance, defined observation in a recent article for the handbook *Literatur und Wissen* [Literature and Knowledge] merely as "aktives und zielgerichtetes Wahrnehmen" (Klausnitzer 2013, 241) [active and focused perception]. But this active and focused perception captures only one – albeit important – part of the whole observational process. In order to truly fulfill his/her aspiration to find new knowledge in unexpected places, the observer cannot focus on a given aspect from the outset. Observation has to begin in a state of openness in which the observer does not yet presume to know what he/she should focus on. Observation, before it becomes a concentrated activity, is the state of being prepared to focus on anything that may impose itself on the future observer as worthy of his or her attention. As Jean-Joseph Menuret writes in his entry "Observateur" [Observer] for Diderot and d'Alembert's *Encyclopédie:* "[T]out le [i. e. the observer] frappe, tout l'instruit, tous les résultats lui sont égaux parce qu'il n'en attend point [...]." (Menuret 1751–1780, vol. 11, 310) [Everything strikes the observer, everything instructs him, all the results are equal to him because he does not anticipate any.] As Menuret emphasizes, the observer does not await anything. Menuret's definition does not merely capture an isolated historic understanding of observation. Published in the most influential Enlightenment encyclopedia, and around the middle of the century that crucially determined the legacy of observation, this notion represents our understanding of observation as such. Not surprisingly, this openness to see any new phenomena that may impose themselves on the observer, which Menuret stresses, is also one of the important points in Lorraine Daston and Elizabeth Lunbeck's recent collection of essays, *Histories of Scientific Observation* (2011). Daston and Lunbeck define the "spirit of observation" as the state of being "open to possibilities for new knowledge in the most unexpected places" (Daston and Lunbeck 2011, 8).

However, while this openness to the object of observation seems indeed crucial to "the spirit of observation," this openness itself, I argue, does not define the entirety of observation either. I call this initial stage of observation, *spectatorship*. In the *Encyclopédie* entry "Spectacles," Louis de Jaucourt emphasizes, in quoting Charles Batteux (1713–1780), the unfocused, even distracted character of spectatorship. The spectator is always eager to watch, but he never focuses on anything:

> L'homme [...] est né spectateur; l'appareil de tout l'univers que le Créateur semble étaler pour être vu et admiré, nous le dit assez clairement. Aussi de tous nos sens, n'y en a-t-il point de plus vif, ni qui nous enrichisse d'idées, plus que celui de la vue ; mais plus ce sens est actif, plus il a besoin de changer d'objets : aussitôt qu'il a transmis à l'esprit l'image de ceux qui l'ont *frappé*, son activité le porte à en chercher de nouveaux, et s'il en trouve, il ne manque point de les saisir avidement. (Jaucourt 1751–1780, vol. 15, 446, my emphasis)

> Man [...] is born spectator; the whole universe, which the creator seems to have displayed to be seen and admired, tells us this clearly enough. Moreover, of all our senses, there is none more vivid, none more enriching to our ideas than the sense of vision; but the more this sense is active, the more it needs to change its objects; as soon as it has transmitted to the spirit the image of what *struck* it, its activity directs it to search for new objects, and if it finds any, it will immediately and greedily grasp them. (My emphasis)

Humans are born spectators, says Louis de Jaucourt – they are born to look at the things that the world has to offer. But the spectator does not focus on any given object. Instead, he/she perpetually and distractedly switches his/her gaze to anything that, as Louis de Jaucourt puts it, "strikes" (*frapper*) his/her attention.

While fundamentally distractible, humans' innate spectatorship initiates the visual engagement with the world, which potentially allows for observation. Rétif's hero is called the nocturnal *spectator*, I am tempted to say, and not the "nocturnal *observer*" – although an unauthorized 1789 edition indeed changes the subtitle to "observateur nocturne" (Rétif 1789, see Jacob 1875, 300–301) – because the novel is interested precisely in the conditions and processes that allow for observation to originate.[4]

As we have seen in the first chapter, in Lesage's novel *Le Diable boiteux* the transition from the student's initially unfocused gaze to the isolation of individual tableaux relies on the combination two distinct factors. The first factor is that Lesage's novel concentrates on an already compartmentalized space of clearly separated rooms and houses; the second factor is that the devil's deictic gestures help to focus the attention on one room after the other. In *Les Nuits de Paris*, by contrast, the nocturnal spectator, who is walking alone through the streets of Paris, can rely neither in the same way on a given compartmentalization of space, nor on the guiding help of a devil. Instead, the nocturnal spectator simply walks through the capital until some unexpected visual or acoustic interruption *strikes* (*frapper*) his attention and brings to the surface a describable tableau:

---

4 My distinction between spectator and observer is different from the one that Jonathan Crary constructs at the outset of *Techniques of the Observer*. For Crary, spectatorship suggests a passive seeing of things while observation calls attention to the fact that vision is an activity that happens "in a system of limitations and conventions" (Crary 1990, 6).

- Je n'avais pas fait cinquante pas dans cette dernière [rue], que je fus *frappé* du son d'une voix plaintive [...].« (Rétif 1788, 30, my emphasis) [I had not made fifty steps in this last street when I was *struck* by the sound of a plaintive voice (...). (My emphasis)]
- Je marchais legèrement ét [sic][5] sans bruit, à l'ombre des maisons, comme le Guet. Un bruit sourd *frappe* mon oreille: Je m'approche: [...]. (Rétif 1788, 58, my emphasis) [I walked leisurely and without noise, in the shadow of the houses, like a watchman. A muffled sound *struck* my ear: I approached (...). (My emphasis)]
- J'alais dans le quartier, pour tâcher d'avoir de nouveaux renseignemens, quand au coin de la rue Traversière, une singularité me *frappa*: [...]. (Rétif 1788, 199–200, my emphasis) [I went into the neighborhood to try to receive new information, when, at the corner of the Rue Traversière a singularity *struck* me: (...). (My emphasis)]

It is important to highlight that the spectator (and potential observer) comes to his objects through such "scènes frappantes" [striking scenes] (Rétif 1788, 230–231) because this decisively distinguishes him in the discourse of the eighteenth century from the experimenter. The experimenter, like the spectator and the observer, engages with the world through its visuality. In contrast to the spectator and the observer, however, the experimenter always already knows the object on which he/she will focus. The experimenter is not interested in being surprised by "striking scenes." In the *Encyclopédie* entry "Observateur," Jean-Joseph Menuret goes so far as to say that, unlike the observer, the experimenter does not see nature itself but only the results of his own combinations (Menuret 1751–1780, vol. 11, 310). Menuret's differentiation between observation and experimentation in the *Encyclopédie* reflects a long-standing scientific debate about these terms since the early eighteenth century – a debate to which thinkers such as Gottfried Wilhelm Leibniz and Christian Wolff contributed (Klausnitzer 2013, 247, Daston 2011, 85). This debate was crucial to the theory and practice of observation, which was defined to a large extent in contradistinction to experimentation. For much of the eighteenth century and up to the mid-nineteenth century, observation fared typically better than experimentation in scientific debates (Daston 2008, 102). Rétif's *Les Nuits de Paris* is in line with this prevalence of observation

---

5 Rétif's spelling deviates here and in many other instances in the following quotations from modern standard French. Quotations are from the novel's original edition. In the twentieth century, several editions with excerpts of the novel appeared (for a complete list see Rétif 1990, 1347). The 1964 English translation (Rétif 1964) also presents only excerpts of the original 3000-page novel.

over experimentation. The nocturnal spectator repeatedly demonstrates the epistemological limits of the social experimenters that he encounters on his nightly walks and who are willing to see only what they anticipated.

One such encounter with the city's experimenters occurs in the 108th night, when the nocturnal spectator perceives a seemingly poor and unsightly woman crying in the Jardin du Luxembourg. As he addresses her, he learns that the woman was deliberately placed in the park by a group of three people who wanted to study the public's reaction to the woman. More precisely, the 'experimenters' made a bet with a young and beautiful woman that, when dressed as old, poor, and ugly, she could spend the busiest hours of the day – that is between three in the afternoon and ten at night – in the park without anyone caring to ask her about her sorrows. Determining thus the object of their study, the experimenters differ from the nocturnal spectator, who runs into this scene by chance. To be sure, the nocturnal spectator goes into the Jardin du Luxembourg because he hopes to find many people there. And in that sense, he has some object in view. He is even annoyed to get there only by nine o'clock, as this park does not remain frequented as long at night as other parks in the city (Rétif 1788, 1185). But beyond that, the nocturnal spectator has no specific goal in mind. He comes to the park as a spectator, intent on watching anything that may arouse his curiosity, but he does not know what exactly he is looking for.

When the experimenters in the Jardin du Luxembourg learn that the person who made them lose their bet by addressing the disguised woman was in fact the "spectateur nocturne," which is how the nocturnal spectator introduces himself to the company, they angrily exclaim that they should have excluded this Parisian character, of whom they have already heard, from their wager. We may easily gather that the reason for this exemption is that the nocturnal spectator somehow behaves strangely, and that he therefore interferes with the framework that governs the experiment. It makes sense for an experimenter to set up such a rule of exemption. For the experimenter necessarily already starts with assumptions about the object he/she wants to study, and his/her task consists in assuring that the actual setting when performing the experiment reflects these previous assumptions. The spectator and observer, by contrast, is characterized precisely by the fact that he/she does not limit the objects of his/her potential observations from the outset and is always open to being surprised. In fact, the whole point of the nocturnal spectator's walks is to encounter "striking scenes" that he would not have expected.

There is another important scene in *Les Nuits de Paris* that emphasizes this contrast between observer and experimenter, and in this scene, the experimental setting is much further developed. In the 73rd night, the nocturnal spectator is led into a brothel. The brothel is run by a group of people who are interested in studying the functioning of prostitution and to find out about its "mille petits de-tails qui conduisaient à connaître la verité" (Rétif 1788, 818 – 819) [thousand little details that help to know the truth]. As in the scene with the crying woman in the Jardin du Luxembourg, the experimenters and the observer react very differently once they realize that the object of their respective study is a scholar of the Pa-risian nights as well. For whereas the experimenters angrily chase the nocturnal spectator away from the brothel – "N'y revenez plus!" (Rétif 1788, 821) [Do not come back here!] – the nocturnal spectator contently asserts: "J'ai payé pour voir ét j'ai vu." (Rétif 1788, 821) [I have paid to see, and I have seen.] Whereas the experimenters, who cannot tolerate any interference with their experimental settings, have lost a night's worth of work, the observer, who is open to seeing anything, including other forms of observation and experimentation that turn himself into an object of study, has added just another scene to his list of curious encounters in the French capital. The significance of this scene in the brothel for the entire novel is also reflected by the fact that it is depicted in the frontispiece of the fourth volume of *Les Nuits de Paris* (Fig. 6).

The first necessary step toward the disclosure of stories through observation is spectatorship, i. e. the openness to attend to whatever appears as curious. This necessary beginning with spectatorship is epitomized in the narrator's an-nouncement in the novel's preface that everything that is told in the novel "peut exciter la curiosité" (Rétif 1788, 2) [can excite curiosity]. Claiming this, the narrator does not merely advertise his book to its potential readers, but also points to the novel's narrative preconditions. For the narrator really can only begin observing and telling that which first imposes itself as curious on the spectator.

Observation

Contrary to the novel's experimenters, who are concerned with pre-defined ob-jects of study, the nocturnal spectator encounters the objects of his investigations as curious surprises. They appear unexpected. The importance of the unexpected character of all initial visual encounters lies in the fact that this very trait neces-sitates an initial concentration on describable, tableau-like, momentary scenes – "Tableaux nocturnes" (Rétif 1788, 2) [nocturnal tableaux], as the preface puts it. Not knowing the object of his observations, the observer has to start with a ta-bleau. He cannot start by watching a narratable sequence of events because this would presuppose the prior focus on a given particular scene. Instead, the yet unfocused or otherwise engaged observer relies on curious, incidental, ta-bleau-like scenes as the beginning of his observations.

Figure 6: The nocturnal spectator in the brothel, under surveillance by social experimenters. Frontispiece by Louis Binet to the fourth volume of the original edition of *Les Nuits de Paris* (Rétif 1788). Source gallica.bnf.fr / BnF.

In the twentieth night, for instance, when still engaged in research related to a previous incident, the nocturnal spectator's attention is suddenly drawn to a new tableau. I already cited the beginning of this scene above as one of the novel's "striking" interruptions:

J'allais dans le quartier, pour tâcher d'avoir de nouveaux renseignemens, quand au coin de la rue Traversière, une singularité me frappa : Un Homme, habillé précisément comme moi, ét de mon âge, était occupé à regarder deux jolies Personnes, dans le comptoir d'une boutique de soierie. Je me mis à l'observer [...]. (Rétif 1788, 199–200)

I went into the neighborhood to try to receive new information, when, at the corner of the Rue Traversière a singularity struck me: a man, dressed exactly like me, and of my age, was busy regarding two pretty women in the showroom of a silk shop. I started to observe him [...].

This scene provides us with a very insightful example of the novel's narratological agenda: it allows us to appreciate not only the different steps in the observational procedure (which leads from the seeing of a striking image to the focused watching of an extended sequence), but also the way in which Rétif turns the analysis of this very procedure into the novel's most imminent concern. Here, as so often in the novel, a singularity strikes the observer ("une singularité me frappa"), which allows for the insertion of a new description in the narrative flow. It is a singularity both in the sense that it stands out as remarkable and in that it is just a *single* thing: the still image of a man looking into a shop window. It is a still image, and as such it can be rendered in description. But it is, moreover, a very peculiar singularity that strikes the observer at the moment that he *traverses* such a self-reflexive place as is the Rue *Traversière*, because the strange and singular man is not only dressed exactly like the nocturnal spectator and of the same age as the latter.[6] Looking at two girls in a shop, the man is also occupied in a very similar, visual activity. Finally, the boutique into which this alter ego of the nocturnal spectator is looking is one of "soierie." In this word "soierie," we may hear just as much the silk (*soie*) as the self (*soi*). Building on the well-known etymological origin of the word "text" in the Latin verb "texere" [to weave], the scene suggests that the *fabric* (silk, *soie*) of the novel's observations is little else but the self (*soi*). What strikes the nocturnal spectator is a singularity in precisely the sense that it is a *doubling* of the nocturnal spectator himself.[7] The nocturnal spectator watches a man who resembles him and who, as the homophony of silk and self in French suggests, is in turn looking at himself through looking at others. Here as in many other cases in *Les Nuits de Paris*, the objects and persons that the nocturnal spectator encounters primarily serve to

---

6 The importance of the dress is reflected in the engravings for the novel, in which the nocturnal spectator is consistently recognizable by his long overcoat and his wide-brimmed hat.

7 The observed man resembles the nocturnal spectator, moreover, in that he is, as we learn later on, an unhappy lover. The nocturnal spectator himself starts his nightly walks originally not with the intention to be an observer, but to commemorate his unhappy loves (Rétif 1788, 8–9).

reflect on the procedure of the nocturnal spectator's observations. *Les Nuits de Paris* is therefore not only a novel *of* observation but also one *about* observation. I will return to this point later on as I continue to discuss what happens to the man in front of the shop window.

Struck by the curious encounter with a man who resembles him so closely, the nocturnal spectator starts to focus on this scene. "Je me mis à l'observer" [I began to observe him], it says in the passage quoted above. However, this process of observation relies already on the prior unfocused gaze of the spectator, which alone could discover scenes worthy of sustained watching. While Rétif's narrator uses the term "observation" for the extended watching, I reserve the term observation for the *combination* of the initial seeing of tableaux and the subsequent sustained watching of chains of events. To the extent that observation is an integral process of producing new knowledge about the world, the nocturnal spectator's extended watching relies on his prior unfocused gaze, which alone makes it possible to discover new images that exist outside of the narrator's present field of knowledge and attention.

As the narrator explains at the outset of the novel, he describes only what he sees: he is "le Hibou-Spectateur, qui ne decrit que ce qu'il a vu" (Rétif 1788, 1) [the owl-spectator who describes only what he has seen]. The nocturnal spectator's seeing consists, however, of two distinct subsequent stages. These two stages of seeing in *Les Nuits de Paris* are, first, the contingent encounter with a striking, describable tableau during the nightly walks through Paris, and second, the subsequent determined and sustained recording of these initially perceived tableaux as they begin to develop into a chain of actions. The initial contingent encounter with a curious moment (the moment in which a new description interrupts the narrative flow) is transformed into the sustained watching of ongoing actions (in which the initial description is extended into the narration). The nocturnal spectator follows the people he has initially noticed through Paris and watches their behavior and interactions until he is informed about their identity and intentions. And this shift from image to sequence is formally represented by the shift from description to narration. The French critic Jean Varloot perfectly captures the nocturnal spectator's visual activities in saying: "le 'hibou' [the owl, i.e. the nocturnal spectator] cerne une image et filme" (Varloot 1987, 12) [the owl sketches an image and films]. However, Varloot fails to expound on the fundamental rupture that exists between the initial determination of the still image and the subsequent continuous watching implicit in what Varloot calls filming. The sequences of events that constitute the novel's stories are mainly disclosed through the sustained recording; but this recording would not be possible if the nocturnal spectator had not previously seen a striking tableau. Neither pure spectatorship nor the pure recording of the sequences alone can produce the curious observations that constitute the material of the novel's stories.

Storytelling

So far, I have discussed only the way in which the nocturnal spectator's observation, which proceeds from the moment of being struck by an image to the sustained watching of a sequence of events, is translated in Rétif's novel into the transition from descriptive to narrative passages. The representation of a visual reality, the novel makes clear, cannot be a matter of description alone, as traditional accounts of literary realism would have it. Quite to the contrary, the disclosure of the world through observation relies on the combination of description and narration. The novel thus performs for the reader a process of perception that gathers new information through the combined seeing of images and sequences, and it explicitly thematizes this process as one of perception by showing these perceptions as the actual perceptions of an intradiegetic character and narrator (the nocturnal spectator). The reader, in other words, is made to see the world as does the nocturnal spectator. However, in this reconstruction of the narrative and perceptual mode of Rétif's novel, literary representation appears to be secondary, merely copying the pre-given visual experience of (non-literary) observation. But the relation between the form of literary representation (storytelling) and observation in *Les Nuits de Paris* is much more complex than that. The novel perpetually stresses that its observations rely also on a framework of storytelling that gives shape to them. More precisely, it is hard to determine whether observation precedes, enables, and determines the form and content of storytelling or whether, conversely, storytelling precedes, enables, and determines the form and content of observation. For *Les Nuits de Paris* does not simply offer observations made in Paris. Instead, it presents the report of the narrator's nightly storytelling of his nocturnal observations to a melancholic marquise whom he meets early on in the novel. The observations that the nocturnal spectator undertakes during the nightly walks through the French capital are thus, from the beginning, part of a framework of storytelling, and this framework of storytelling determines the form of the novel just as much as the fact that the novel offers an account of observations. The narrator explicitly stresses that the narrative form of the book is due to the fact that he gave an account of his observations to the marquise: "Il [the nocturnal spectator] a donné à cet Ouvrage la forme animée d'un recit ; parce-qu'effectivement, il a rendu-compte à une Femme de tout ce qu'il voyait." (Rétif 1788, 2) [He gave this work the animated form of a narrative, for, in fact, he told a woman everything that he saw.]

In this combination of the observations made in Paris with the insistence on recurring acts of storytelling, *Les Nuits de Paris* appears as something like a hybrid
between Sébastien Mercier's popular *Tableau de Paris* (1781) and the *Arabian
Nights* (first French and European translation 1704–1717) – two works out of
whose shadow Rétif's novel has never really emerged.[8] Not coincidentally, the
nocturnal spectator notes at the very outset of the novel that he spent 1001
nights walking through Paris: "Dans le cours de vingt années, c'est à dire, depuis
1767, que l'Auteur est Spectateur nocturne, il a observé pendant 1001 nuits, ce qui
se passe dans les rues de la Capitale [...]." (Rétif 1788, 1) [During the course of
twenty years, that is since 1767, the author has been a nocturnal spectator and
he has observed during 1001 nights what happens in the streets of the capital.]
Through the reference to the 1001 Arabian nights, the observational project is
grounded from the beginning in a framework of storytelling.

But the relation between observation and storytelling in Rétif's novel is even
more complex. For the acts of storytelling themselves are repeatedly situated
within observed sequences. The nocturnal spectator observes people summariz-
ing and explaining the observed sequences of events. Even his own telling of the
stories to the marquise whom he visits at the end of almost every night, are, to a
certain extent, still part of his observations because the marquise was the very
first person whom he encountered on his nightly walks (Rétif 1788, 2). The
later storytelling is thus, as it were, still part of the initial scene of observation.
Neither one of the two – observation and storytelling – exists independent of the
other. The point is thus not so much to debunk the empirical practice of observation as structured by the rules of storytelling. What Rétif does, instead, is to
show how observation and storytelling depend on each other.

This interlocking of observation and storytelling can also be studied in the
scene with the man who so curiously resembles the nocturnal spectator. The nocturnal spectator stands still long enough to see that the mysterious man who is
observing two girls through a shop window throws a letter into the shop. The
man subsequently walks up and down in front of the shop, until he is finally
seized by two young men and carried into the shop's back room. These last
events attract the attention of the neighboring crowd. Together with the nocturnal spectator, the crowd gathers around the shop and watches the scene inside.
Inside the shop, the captive is sitting encircled by the shopkeeper's family, and
he is pressed to confess his motive for writing the letters. Finally, the family's fa-

---

8 Other important texts that influence the general project of *Les Nuits de Paris* include Addison
and Steele's journal *The Spectator* (1711–1712, published in numerous new editions throughout
the eighteenth century) as well as Marivaux's *Le Spectateur français* (1721–1724) (Varloot 1987,
24; for a comprehensive study of the literary context of *Les Nuits de Paris*, see Barr 2012).

ther arrives and chases the crowd away in order to interrogate the captive in private. Together with the neighboring crowd, the nocturnal spectator witnesses the decisive scene of interrogation from afar. The continued observations thus lead into an observed scene of interrogation that captures their meaning – a meaning, albeit, that is not initially clear to the nocturnal spectator, who cannot overhear the conversation.

Les Nuits de Paris functions in this equilibrium of completely observed stories that attest to the necessity of their telling and explanation within the observed sequences: the act of storytelling is part of the observation and yet distinct from it. We can witness the movement toward storytelling within the observation several times more within the sequence about the man in the silk shop. As the interrogated man leaves the house through a back door, the nocturnal spectator joins him: "Il était onze heure: Je le joignis: ––Jeunehomme (lui dis-je), voila une singulière avanture ! Si je puis te servir en quelque chose, dis le moi– ?" (Rétif 1788, 201) [It was eleven o'clock. I joined him. Young man, I said to him, this is a singular adventure. If I could be of any service to you, would you tell me?] The direct address of the observed man marks the beginning of a third stage in this sequence – after the initial arousal of curiosity by the striking tableau and the subsequent continued watching.[9] However, the ensuing conversation, which integrates the previously observed events into a coherent story, remains in a way still part of the nocturnal spectator's observations because the conversation is part of what the nocturnal spectator subsequently tells the marquise – to whom "il a rendu-compte [...] de tout ce qu'il voyait" (Rétif 1788, 2) [he reported everything he saw].

It turns out that, some time ago, the man had fallen in love with one of the shopkeeper's daughters. The most striking features of the daughter, the man says, consist "d'un regard enchanteur, ét d'un pied mignon" (Rétif 1788, 202) [of enchanting eyes and an adorable little foot]. Ever since, the man has followed her. When merely watching the young woman no longer satisfied his desires, he started writing letters to her – thus supplementing, as it were, his deficient observations with accounts of his observations: "Enfin, la vue ne suffisant plus, j'ai écrit les lettres les plus tendres." (Rétif 1788, 201) [Finally, when watching did not suffice any longer, I wrote the most tender letters.] However, as the man did not hope to have any success in his pursuits, he remained hidden from her, until this

---

9 The beginning of this third phase is clearly marked by the mentioning of the time ("Il était onze heure : Je le joignis"). The narrator had mentioned the time of day only once before, at the very beginning of this episode: "Le lendemain, à neuf-heures-ét-démie, je quittai ma chambre ét mon travail." (Rétif 1788, 199) [The following day, at half past nine, I left my room and my work.]

evening, when he was found out by the girl's family and forced to confess his love.

Despite the man's modest social standing, the shopkeeper is initially inclined to accept him as suitor for his daughter. But the shopkeeper demands of the man that he prove the constancy of his love. The shopkeeper therefore asks the man to wait for one year before his next visit. The nocturnal spectator congratulates the *malade-d'amour* on this promising news. But at this point the lovesick man confesses that he is, in fact, already married and thus devoid of all hope to gain the shopkeeper's trust.

After the lovesick young man has finished his story, the nocturnal spectator goes to the marquise to whom he tells the night's events. The move from observation to storytelling appears thus four times within in the sequence of the lovesick man: in his letters to the shopkeeper's daughter, in the interrogation by the shopkeeper, in the conversation between the lovesick man and the nocturnal spectator, and, finally, in the nocturnal spectator's storytelling to the marquise. We should remember here also that the marquise herself is the first person the nocturnal spectator observed in the whole novel. In telling the story of the lovesick man to the marquise, the nocturnal spectator is thus to a certain extent still observing (the marquise).

The four stages of observation and storytelling relate to each other like Russian matryoshka dolls, each encapsulating the previous scenes of storytelling and observation. The shopkeeper's daughter is beautiful mainly because of her eyes (and, of course, because of her feet, but that's a given in Rétif's world of foot and shoe fetishism; see Wyngaard 2012). The lovesick man thus observes a beautiful 'observer,' as it were, and he writes letters to her in which he presumably speaks about both his and her 'gaze.' The lovesick man in turn is observed by the girl's family members, who subsequently interrogate him. Finally, the nocturnal spectator questions the man about the observed interrogation. Time and again, the transition from observation to storytelling is itself the very object of observation. The observed reality strives unavoidably toward its telling, but this movement from observation to storytelling is in turn fundamentally observed. Only to the extent that there is an observed reality can there be storytelling, but conversely, the observed reality becomes thematic only if it is told (in the form of a story). *Les Nuits de Paris*, in other words, may perform a kind of seeing that is open-ended (ready always to be drawn to curious new phenomena), but it also makes clear that this performance of the openness of observation happens within a relatively closed framework of storytelling. Stories, I submit here, possess a beginning, middle, and an end and thus show some degree of coherence – which in turn determines what can occur in them.

The reliance of the observational regime on a traditional framework of storytelling helps also to explain a curious aspect of Rétif's novel that I have so far ignored. For the account of the nights spent in Paris is, in the first volume of the novel, repeatedly interrupted by a lengthy tale about the Greek figure Epimenides. According to Rétif's version of the ancient myth, Epimenides slept for 75 years in a cave in Mount Ida on Crete before returning to his people. In *Les Nuits de Paris*, the nocturnal spectator tells the story of Epimenides to the marquise, alongside the observations of the events taking place in Paris. The combination of the ancient myth with the contemporary observations seems at first confounding, breaking with modern expectations of artistic unity and coherence. A closer look, however, reveals that the telling of the story of Epimenides is indeed crucial to the project of Rétif's novel. One thing to consider here is that the story of Epimenides mirrors the novel's general concern to tell an interesting and new reality: the narrator is reminded of the story of Epimenides as he sees Paris grow and begins to wonder how Paris would strike one if one were to return to it in a hundred years (Rétif 1788, 48). But still more important is that, to the extent that the observational processes in Rétif's novel always rely on an act of storytelling, Rétif's account of observations in general seems to depend as well on an accompanying traditional fable. The myth of Epimenides fulfills precisely this function: it provides the traditional narrative alongside which the new form of observation-based storytelling can flourish. This economy of storytelling may appear strange from today's perspective, but it seems much less strange if we consider Rétif's attempt to construct a full narrative reality from procedures of observation as a still relatively new endeavor at his time. Rétif novel shows how an observational literary regime is born from within the tradition of storytelling.

The nuanced elaboration of the different stages in the process of observation and the unfolding of the dialectical relation between observation and storytelling are, I argue, the central concerns of *Les Nuits de Paris*. To be sure, there is a grand anthropological, moralist, and pedagogical program that the nocturnal spectator presents at the outset of the novel. As he tells us, he errs through the city to "know man": "J'errais seul, pour connaître l'Homme..." (Rétif 1788, 3) [I was erring alone to know man...]. He sees what no one has seen (Rétif 1788, 5) and speaks about what he has seen to astound and instruct the peaceful citizens (Rétif 1788, 3) and their still innocent children (Rétif 1788, 4). He wants to tell them secrets of the horrible features of vice and crime and injustice, so that they rejoice in the luck of being far from the frightful dark side of life in the city. However, the list of things that the nocturnal spectator proclaims to present lacks any real focus and is practically infinite – Rétif 'ends' it in fact with no fewer than six dots.

Vous y verrez des Filles, des Femmes, des Catins, des Espions, des Joueurs, des Escrocs, des Voleurs : Vous y verrez des actions secrettes ét genereuses, qui relèvent l'Humanité, qui la rapprochent de son divin Auteur : Vous y trouverez de la morale, de la philosophie...... (Rétif 1788, 6–7)

You will see here girls, women, prostitutes, spies, gamblers, swindlers, thieves; you will see secret and generous acts that lift up humanity and move it closer to its divine maker; you will find morality, philosophy......

Yes, there are certain themes that play an important role in *Les Nuits de Paris:* crime, male erotic desire, and the pitfalls of female sexuality, to name but three especially prominent ones. But these are topics about which Rétif writes in almost all his works, and their treatment in *Les Nuits de Paris* is probably not the most interesting and surely not the most shocking. It is no wonder, for instance, that Amy S. Wyngaard, in her well-researched new monograph on sexuality and pornography in Rétif's works, hardly ever mentions *Les Nuits de Paris*. Although sexuality is the topic of every other story in *Les Nuits de Paris*, the novel's main interest lies elsewhere. This is not to say that it would not be worthwhile also to explore the content of Rétif's musings in some more detail. For instance, William F. Edmiston's analysis of the nocturnal spectator's authoritarian politics and their relation to the changing understanding of the state in eighteenth-century France certainly offers one fruitful avenue of critical inquiry. But reading the 3000 pages of *Les Nuits de Paris*, it is hard to escape the impression that this novel is, most fundamentally, not about any of the individual objects it displays, but about the analysis of its own procedures of spectatorship, observation, and storytelling. The questions that the novel sets out to answer are, first, how one tells the story of the observation of reality and, second, to which extent this act of storytelling shapes the observations themselves. In the sequence about the lovesick man as in so many other sequences of the novel, the attention to the procedure of observing and telling what one observes completely overrides other topics, such as, for instance, love, marriage, and fidelity, which are also discussed.

The main concern of Rétif's novel lies in the figure of the nocturnal spectator himself and in his procedures of observation and storytelling. This strong emphasis on the procedures of the observer distinguishes Rétif's novel *Les Nuits de Paris* from Lesage's novel *Le Diable boiteux*, in which devil and student, generally speaking, assume only a marginal role.

There are thus two crucial differences between *Le Diable boiteux* and *Les Nuits de Paris*. The first difference is that the visual itself is allowed a more prominent position in Rétif's novel, in which entire stories are revealed through observational procedures. In Lesage's novel, by contrast, the presently visible scenes under the roofs of the houses of Madrid serve (outside the framing narrative) only as starting points for the devil's narratives. The second difference is that Rétif's novel also pays much closer attention to the different stages of the process in which we engage with the world as it visually appears to us. In other words, not only the visual itself, but also the analysis of the activity of observation plays a more prominent role in *Les Nuits de Paris*.

A comparison of the main illustrations of Rétif's and Lesage's novels confirms this second difference between the two texts.[10] While Lesage's devil and student are almost hidden in the top left corner of the engraving by Dubercelle that I analyzed in the first chapter (see Fig. 2), Rétif's nocturnal spectator occupies the foreground in the frontispiece to the first volume of *Les Nuits de Paris* (see Fig. 7). Moreover, the figure of the nocturnal spectator is doubled at least twice in the engraving, once in the owl that rests on the hat of the nocturnal spectator (*le Hibou-Spectateur* [the owl spectator], as he is also called) and a second time in the owl that flies in the sky above the houses. Through these mirrorings, the nocturnal spectator himself becomes an object of his own observations. The engraving, in other words, draws our attention to the fact that the nocturnal spectator's observational procedures are not only the condition of storytelling in *Les Nuits de Paris*, but also the main object of storytelling.

At the moment when observation truly comes into being, one might be tempted to say here, it also includes some form of second-order observation: it observes (also) itself. However, one should be careful not to generalize on the basis of Rétif's novel. Literary observation performs, in my definition, primarily a first-order process of perception, and texts that employ observations do not necessarily also have to reflect on these procedures (the observation at the very beginning of *Le Diable boiteux* is a case in point).

Rétif's *Les Nuits de Paris* offers us a first moment of closure in this study. Here, we have a novel that not only heavily relies on the literary device of observation, but that also analyzes the epistemological and literary conditions and implications of the kind of seeing performed by literary observation in admirable clarity. Rétif thus offers a model of what observation is – both as an epistemological and as a literary practice. However, the analysis of the procedure of literary observation in this book would remain incomplete without tending also to the various texts in which the failure of observation is a central concern. More than just abstractly showing us that literary observations are not as uncomplicated as they may appear in *Les Nuits de Paris*, texts about the failure of observation

---

10 Rétif, at least, is known to have closely collaborated with his illustrator, Louis Binet (Barr 2012, 35–36).

raise our awareness for the broad range of cultural contexts that enable – or inhibit – literary observations. In the following chapter, I will present readings of three texts that show us three different ways (and three different contexts) in which literary observation fails.

Figure 7: "Le Hibou Spectateur." Frontispiece by Louis Binet to the first volume of the original edition of *Les Nuits de Paris* (Rétif 1788) Source gallica.bnf.fr / BnF.

Chapter 4: Failing Observations

Alain-René Lesage's *Le Diable boiteux* inaugurates a literary tradition in which the construction of the stories is based on procedures of seeing the world. However, the precise mode of seeing that is presented in Lesage's novel is ultimately very limited and falls short of observation. For while the flying devil is able to show Madrid from above and to make the roofs of the city's houses disappear, the revealed scenes consist merely in tableaux of isolated moments in time. The engagement with the visual world only renders describable tableaux, not narratable sequences. As the devil explains, the true meaning of the revealed tableaux can be understood only by learning about their backstories, and these background stories cannot be deduced from the tableaux.

Although not itself a novel of observation proper, Lesage's *Le Diable boiteux* became an important reference point of a rich European tradition of narrative fiction that develops observational procedures to construct a full visual reality, consisting of both static images and dynamic sequences. This tradition includes stories of people walking through the streets of Paris and London, as in Rétif's *Les Nuits de Paris* and Daniel Defoe's *A Journal of the Plague Year*; of elderly men watching from their windows the events below, as in E.T.A. Hoffmann's "Des Vetters Eckfenster" [My Cousin's Corner Window] or Wilhelm Raabe's *Die Chronik der Sperlingsgasse* [The Chronicle of Sparrow Lane]; and of detectives reconstructing crimes based on their detailed scrutiny of the environment, as in Edgar Allan Poe's "The Murders in the Rue Morgue" or Arthur Conan Doyle's Sherlock Holmes stories. These stories construct both description and narration as procedures of seeing. But the technique of observation that I discuss in this book extends far beyond these explicitly observational texts in which the entire act of storytelling is mediated by a homodiegetic narrator who looks at the world. Literary observations – in my sense of this term – occur in all those short and limited scenes in which we newly encounter (through description) a character (or object) and then see how this character (or object) is set in motion. Observations, in other words, do not have to be explicitly marked and thematized as acts of seeing; they can also be simply performed by the text (recall once more the opening observation of *Le Diable boiteux*, where the heterodiegetic narrator simply tells us that Don Cléofas emerges from the attic and then has him run over the roofs of the houses). In those cases, however, in which the observations are also thematized as acts of seeing within the text, this thematization often helps clarify something about the process of seeing that is implicit in any literary observation.

https://doi.org/10.1515/9783110594348-111

Les Nuits de Paris is an especially striking case of such an illumination of the process of seeing that defines observation. Indeed, Retif's novel allowed me to discuss the ways in which the literary device of observation also implies an epistemological duality. More precisely, I argued that proper observation – the unprejudiced, knowledge producing registration of the world as it appears to our eyes – consists in the combination of two forms of seeing: the initial undirected gaze of the spectator, which is limited to the perception of describable images, and the sustained and focused watching of sequences. In Rétif's voluminous novel, we see the nocturnal spectator again and again make the transition from the registration of describable tableaux to the sustained watching of narratable sequences.

But while Rétif's ideal procedure importantly gestures toward the possibility of observation to disclose a dynamic visual reality, the feasibility of the transition from the initial openness to be struck by a curious image to the focused recording of a coherent sequence of events is less simple than Rétif suggests. Rétif's nocturnal spectator easily switches from the description of momentary tableaux to the recording of temporally extended sequences. But how natural is this transition? Why will the nocturnal spectator not be perpetually distracted by something else? And even if one image really arrests the nocturnal spectator's attention as he is walking through Paris, why would he not try to 'photograph' the stable image, as it were, instead of 'filming' the ensuing sequence? Finally, what is it that turns the ensuing sequence into something more than a mere series of disconnected images? These and similar questions are neither wholly abstract nor purely contrived. As I will show in this chapter through exemplary interpretations of works by Johann Wolfgang Goethe, Georg Büchner, and Edgar Allan Poe, these questions are of formative importance for many literary narratives from the decades around the publication of *Les Nuits de Paris*. Moreover, these questions emerge from a wide range of discourses and technologies of seeing (from aesthetics through morphology to urban spectatorship) as well as of literary traditions (from the *Sturm und Drang* to early nineteenth-century realism).

Before turning to Poe's story "The Man of the Crowd," I first will focus on Goethe's epistolary novel *Die Leiden des jungen Werthers*[1] (first version 1774, revised version 1787) and Büchner's novella *Lenz* (written around 1835, published posthumously in 1839). Goethe's and Büchner's texts offer two complementary versions of the problem of the transition from image to sequence. Werther, in my reading, is first of all the ideal *absorbed* spectator of the eighteenth century,

---

1 My reading of *Werther* draws, in part, on an earlier article of mine (Wagner 2012).

in the sense in which Michael Fried described this type of beholder in his study *Absorption and Theatricality: Painting and Beholder in the Age of Diderot* (1980). Like the contemporary viewer of artworks, who was "stopped and held, sometimes for hours at a stretch if contemporary testimony is believed, in front of the painting" (Fried 1988, 132), Werther dwells in the contemplation of still scenes in nature. Contemplating these natural tableaux, Werther, moreover, imagines his entrance and immersion into them. Resembling again the contemporary visitors of galleries who enjoyed the "fiction of [their] physical presence within the painting" (Fried 1988, 132), Werther imagines his presence in the scenes in nature that he beholds. Werther's way to relate to the visual, in other words, is to be completely taken by what he sees – to the extent that he imagines becoming part of the scene and assuming the position of the object of his own contemplation. Importantly, the stability of the registered tableaux is the condition of possibility for Werther's imagined entrance into the scene. Only if the tableau does not change is it possible to imagine one's entrance into it. Unable to see more than individual images, Werther is incapable of becoming an observer. Werther fails to become an observer because he clings to individual tableaux of persons and objects and remains unwilling to account for their change over time. Werther's shortcomings as an observer also account for the fact that his letters, which capture his encounters with the world, do not provide full narratives and that the novel eventually has to transition from the mere reliance on these letters to the report of an extra-diegetic narrator.

Büchner's Lenz, by contrast, embraces the instability of the perceived images. He famously emphasizes this instability in the novella's central conversation on art (*Kunstgespräch*). Instead of clinging to any single image, Lenz affirms a world that provides ever new images, one after the other. However, Lenz's affirmation of the ephemeral nature of the perceived tableaux also falls short of observation. Lenz remains a spectator of series of fragmentary perceptions, not an observer of coherent sequences. Büchner's very prose embodies this failure of observation, disintegrating into long paratactic chains that reflect Lenz's fragmentary perception.

To be perfectly clear, I do not claim that either Goethe's or Büchner's treatment of observational procedures was developed in direct response to Rétif's novel – in any case, Goethe's *Werther* was published before *Les Nuits de Paris*. Instead, I suggest that the decades around the publication of Rétif's novel are marked by a heightened general awareness of procedures of observation in science, politics, and the arts. In the case of both Goethe and Büchner, this general awareness of the intricacies of observation is intensified by the authors' scientific work. To be more precise, I argue that Büchner's discussions of observational procedures in literature can be understood as analogous to the theories of scientific observation that he develops as a student of comparative anatomy. To a lesser extent, such a parallel exists also between Goethe's narrative and his work on the morphology of plants many decades later. But whether or not one agrees that Goethe's and Büchner's texts are explicitly and intentionally about (scientific and literary) procedures of observation, their texts help us fathom the conditions that lead to the failure of literary observations. As simple as observational sequences like those we see in *Les Nuits de Paris* may seem, they are nearly impossible in a wide range of cultural and literary contexts.

Die Leiden des jungen Werthers

To understand the complex problems that Werther faces as an observer, we first have to disabuse ourselves of a cliché of literary history according to which Werther is primarily a lover. Above being a lover, Werther is simply someone who *sees* the world. It is his particular form of seeing that subsequently turns him into a lover of the perceived objects. Love is not merely a way for his visual engagement with the world to become thematic in the novel. Love is the consequence of a particular kind of seeing; it is the expression of Werther's attempt to become part of the scenes that he perceives.

The prominence of the theme of visual perception is already palpable in the novel's organization around elevated viewpoints – a structural element on which critics, to my knowledge, have never commented. Many of the most prominent places in Goethe's novel are characterized by the fact that they offer wide views over the surrounding area. This is particularly foregrounded in the first letters of the novel. Already in the opening letter, Werther mentions a garden where he spends many hours and which is set on a hill (Goethe 2006, 13). In the third and again in the fifth letter, Werther describes a well above which he sits many hours watching women carry away water (Goethe 2006, 17). In the ninth letter, Werther introduced the reader to his favorite place, the hamlet Wahlheim, which is situated, again, on a hill, so that he can overlook the whole valley (Goethe 2006, 27). This *Wahlheim* – literally, Werther's "chosen home" – becomes the supreme point of surveillance and produces many of the scenes of the novel: "es ist wieder Wahlheim, und immer Wahlheim das diese Seltenheiten hervorbringt" (Goethe 2006, 33) [It is Wahlheim once more—always Wahlheim—which produces these wonders (Goethe 1854, 257)]. On his walks around Wahlheim, Werther repeatedly watches Lotte's house from afar, even before he has met her.[2] Again, when Werther and Lotte first come close to each other in the famous moment at the country ball after the thunderstorm (twelfth letter), they sit at a window that allows them to overlook the area: "Wir traten an's Fenster. [...] Sie [Lotte] stand auf ihren Ellenbogen gestützt; ihr Blick durchdrang die ganze Gegend [...]." – (Goethe 2006, 53) [We went to the window [...]. Charlotte leaned forward upon her arm; her eyes wandered over the scene [...]. (Goethe 1854, 265)] Finally, after his suicide, Werther is found lying on the floor, near the window: "Er lag gegen das Fenster entkräftet auf dem Rücken [...]." (Goethe 2006, 265) [He was found, lying on his back, near the window. (Goethe 1854, 354)] As this abundance of lookout points in the novel makes clear, Werther systematically searches for an approach to the world that is fundamentally visual.

The fact that many readers have ignored the large number vista points in the novel – and along with it, Werther's 'visual identity' (his identity as someone who sees) – may also be, curiously enough, an effect of Werther's own desperate attempts to distract from them. While Werther relies on these vista points and the outside perspective that they offer for his descriptions, his goal is to enter the scenes from these vista points in a moment of total immersion. Werther does not want to remain an outside spectator. He wants to be part of the scene that he encounters as a spectator. There is a marked contrast between the initial emphasis on elevated vista points in the novel and the ideal of immersion foregrounded in the longer descriptive passages. To get a sense of Werther's attempt to efface his own perspective, consider here, for instance, the odd way in which Werther's detailed description of the valley in his second letter seems to proceed both top-down (from the sun to the ground) *and* bottom-up (from among the weeds on the ground, where Werther is lying):

> Wenn das liebe Thal um mich dampft, und die hohe Sonne an der Oberfläche der undurchdringlichen Finsterniß meines Waldes ruht, und nur einzelne Strahlen sich in das innere Heiligthum stehlen, ich dann im hohen Grase am fallenden Bache liege, und näher an der Erde tausend mannichfaltige Gräschen mir merkwürdig werden; wenn ich das Wimmeln der kleinen Welt zwischen Halmen, die unzähligen, unergründlichen Gestalten der Würmchen, der Mückchen näher an meinem Herzen fühle, und fühle die Gegenwart des Allmächtigen der uns nach seinem Bilde schuf, das Wehen des Allliebenden, der uns in ewiger

---

2 "Hätt' ich gedacht, als ich mir Wahlheim zum Zwecke meiner Spaziergänge wählte, daß es so nahe am Himmel läge! Wie oft habe ich das Jagdhaus, das nun alle meine Wünsche einschließt, auf meinen weiten Wanderungen, bald vom Berge, bald von der Ebne über den Fluß gesehen!" (Goethe 2006, 57) [Little did I imagine when I selected Wahlheim for my pedestrian excursions, that all heaven lay so near it. How often in my wanderings from the hill-side or from the meadows across the river, had I beheld this hunting-lodge, which now contains within it all the joy of my heart. (Goethe 1854, 266)]

Wonne schwebend trägt und erhält; mein Freund! wenn's dann um meine Augen dämmert, und die Welt um mich her und der Himmel ganz in meiner Seele ruhn wie die Gestalten einer Geliebten; dann sehne ich mich oft und denke: ach könntest du das wieder ausdrücken, könntest dem Papiere das einhauchen, was so voll, so warm in dir lebt, daß es würde der Spiegel deiner Seele, wie deine Seele ist der Spiegel des unendlichen Gottes! (Goethe 2006, 15)

When the lovely valley teems with vapour around me, and the meridian sun strikes the upper surface of the impenetrable foliage of my trees, and but a few stray gleams steal into the inner sanctuary, then I throw myself down in the tall grass by the trickling stream, and as I lie close to the earth, a thousand unknown plants discover themselves to me. When I hear the buzz of the little world among the stalks, and grow familiar with the countless indescribable forms of the insects and flies, then I feel the presence of the Almighty, who formed us in His own image, and the breath of that universal love which bears and sustains us, as it floats around us in an eternity of bliss; and then, my friend, when darkness overspreads my eyes, and heaven and earth seem to dwell in my soul, and absorbs its power, like the idea of a beloved mistress, then I often long and think: O! that you could describe these conceptions, that you could impress upon paper all that lives so full and warm within you, that it might be the mirror of your soul, as your soul is the mirror of the infinite God! (Goethe 1854, 249)

At first sight, Werther's perspective in this passage is almost the inversion of the perspective he would have from any of the vista points just mentioned. Werther does not stand on any elevated spot, but, instead, lies in the high grass, under the forest's trees, in the valley. However, there is something like a second perspective present in this passage – the perspective from which the description itself proceeds. Curiously, the perspective from which Werther organizes his description is diametrically opposed to the one Werther himself inhabits. While he lies on the ground, the text advances, as it were, from above – from the rays of the sun that first land on the treetops and then transgress into the interior of the forest and continue their ways through the weeds and onto the ground where Werther lies. The tension between the opposed perspectives in this passage, in which the description proceeds both from below and from above, points toward the possibility of the convergence between viewer and viewed. It brings Werther into the center of his own description.

Alternatively, one could argue that there is no perspective at all in this passage. The passage stages the attempt at total immersion through the undoing of perspective – and even through the undoing of seeing. Rather than seeing the world from above, Werther is shut off from (almost) all light under "der undurchdringlichen Finsterniß meines Waldes" [the impenetrable darkness[3] of my trees]. What he perceives around him is described not so much as a product of seeing as

---

3 Boylan translates "foliage" instead of "darkness" (*Finsternis*).

of feeling: "wenn ich das Wimmeln der kleinen Welt zwischen Halmen, die un-
zähligen, unergründlichen Gestalten der Würmchen, der Mückchen näher an
meinem Herzen *fühle*" (my emphasis) [when I *feel*[4] the buzz of the little world
among the stalks, and grow familiar with the countless indescribable forms of
the insects and flies (my emphasis)]. Werther does not try to describe what he
sees; instead, he describes what lives inside him: "ach, könntest du das wieder
ausdrücken, könntest dem Papiere das einhauchen, was so voll, so warm in dir
lebt." [O! that you could describe these conceptions, that you could impress
upon paper all that lives so full and warm within you.] However, this emphasis
on feeling is, in some sense, just another way for Werther to negate his existence
as a seeing subject and to eliminate the distance between himself and the world
that he sees.

Moments in which Werther attempts to place himself simultaneously as subject and object of his descriptions appear in different forms throughout the
novel. One of the most remarkable incidents of a collapsing into one of viewer
and viewed occurs in the sequence of scenes that gives an account of the production of Werther's sole actual work of art. This sequence begins in the letter of May
26 – the same letter in which Werther describes his favorite place, Wahlheim,
which, as we saw, is prominently placed on a hill, allowing for wide views
over the valley. Werther sits alone at a table in front of the church. It is afternoon
and the village is almost empty as nearly everyone is out working in the fields.
Only two little boys are left. Werther sees and describes them as they rest almost
motionlessly on the ground:

> Es war alles im Felde, nur ein Knabe von ohngefähr vier Jahren saß an der Erde und hielt ein
> anderes, etwa halbjähriges, vor ihm zwischen seinen Füßen sitzendes Kind mit beyden Ar-
> men wider seine Brust, so daß er ihm zu einer Art von Sessel diente, und ohngeachtet der
> Munterkeit, womit er aus seinen schwarzen Augen herumschaute, ganz ruhig da saß. (Goethe
> 2006, 27)

> Everybody was in the fields, except a little boy about four years old, who was sitting on the
> ground and held between his knees a child about six months old; he pressed it to his
> bosom with both arms, which thus formed a sort of arm-chair, and notwithstanding the
> liveliness which sparkled in its black eyes, it remained perfectly still. (Goethe 1854, 254)

The pronounced quietude is significant. Werther needs an exceptionally calm
setting in which no one moves to describe – and to begin to love – his objects
of contemplation. In this need for absolutely still objects, he resembles, in a
way, the early photographer, who could take pictures only of that which was

---

4 Boylan translates "hear" instead of "feel" (*fühle*).

able to stand still for the long exposure times necessary for the first cameras. As Walter Benjamin suggests in his "Kleine Geschichte der Photographie" [Short History of Photography], the abundance of cemeteries on photographs taken in the mid-nineteenth century was partly motivated by the search for such quiet environments (Benjamin 1991, 373). Werther's view, like that of the nineteenth-century photographer, is biased: Werther can see and represent only that which remains still. Later on, we will see that the questioning of the possibility of any such motionlessness in nature centrally influences the reflections on observation in *Die Leiden des jungen Werthers* and in *Lenz*.

Werther, pleased by the view of the peaceful children, sits down on a nearby plough and starts drawing the brotherly scene. One hour later his work is finished:

> Ich fügte den nächsten Zaun, ein Scheunenthor und einige gebrochene Wagenräder bey, alles wie es hinter einander stand, und fand nach Verlauf einer Stunde, daß ich eine wohlgeordnete sehr interessante Zeichnung verfertiget hatte, ohne das mindeste von dem meinen hinzuzuthun. (Goethe 2006, 27–29)

> I added the neighbouring hedge, the barn-door, and some broken cart-wheels, just as they happened to lie; and I found in about an hour that I had made a very correct and interesting drawing, without putting in the slightest thing of my own. (Goethe 1854, 254–255)

Werther returns twice to this scene in the following letters, retrospectively constructing an ever more intense absorption in the scene. Thus, in the letter of May 27, he claims to have spent *two* hours in creative contemplation – not one hour, as he previously had said: "Ich saß, ganz in mahlerische Empfindung vertieft, die dir mein gestriges Blatt sehr zerstückt darlegt, auf meinem Pfluge wohl *zwey* Stunden." (Goethe 2006, 31, my emphasis) [Absorbed in my artistic contemplations, which I briefly described in my letter of yesterday, I continued sitting on the plough for *two* hours. (Goethe 1854, 256, my emphasis)] Even more striking, in the letter from May 30, Werther confuses the object of his drawing. For while he claims in the two letters before that he sat on a plough while drawing the children, who were sitting on the ground (compare Nisle's illustration, Fig. 8), he now claims to have in fact drawn a plough: "Ein Bauerbursch kam aus einem benachbarten Hause und beschäftigte sich an dem Pfluge, den ich neulich gezeichnet hatte, etwas zurecht zu machen." (Goethe 2006, 35) [A peasant came from an adjoining house, and set to work arranging some part of the same plough which I had lately sketched. (Goethe 1854, 257)][5] Confusing the object of his drawings, Werther, by implication, enters the scene that he has drawn.

---

5 Granted, it is possible that Werther refers here to a drawing that we have not heard of and

DIE LEIDEN DES JUNGEN WERTHER.

I.

Figure 8: Julius Nisle, No. 1 of 12 steel engravings to Goethe's *Die Leiden des jungen Werthers* (Nisle 1840). Source Klassik Stiftung Weimar / Herzogin Anna Amalia Bibliothek / F 3337 (f).

While Werther merely fantasizes of becoming one with the object of his own contemplation in this case, in another series of scenes he actually attempts to replace the object of his description with himself. This latter sequence starts with the scene of Werther's first encounter with Lotte, the object of all his later desires. On his way to a country ball, Werther picks up Lotte from her home. When Werther enters her house, she is still busy preparing supper for her younger siblings and does not notice his intrusion in the domestic scene:

---

which he may have mentioned in a letter that is not included in the collection made for the novel. For, as the editor points out, he did not necessarily publish all of Werther's letters, but only those he was able to find: "Was ich von der Geschichte des armen Werthers nur habe auffinden können, habe ich mit Fleiß gesammlet und lege es euch hier vor [...]." (Goethe 2006, 11) [I have carefully collected whatever I have been able to learn of the story of poor Werther, and here present it to you [...]. (Goethe 1854, 247)]

> Ich ging durch den Hof nach dem wohlgebauten Hause, und da ich die vorliegende Treppen hinaufgestiegen war und in die Thür trat, fiel mir das reizendste Schauspiel in die Augen, das ich je gesehen habe. In dem Vorsaale wimmelten sechs Kinder von eilf zu zwey Jahren um ein Mädchen von schöner Gestalt, mittlerer Größe, die ein simples weißes Kleid, mit blaßrothen Schleifen an Arm und Brust, anhatte. Sie hielt ein schwarzes Brod und schnitt ihren Kleinen rings herum jedem sein Stück nach Proportion ihres Alters und Appetits ab, gab's jedem mit solcher Freundlichkeit und jedes rufte so ungekünstelt sein: Danke! indem es mit den kleinen Händchen lange in die Höhe gereicht hatte, ehe es noch abgeschnitten war, und nun mit seinem Abendbrode vergnügt, entweder wegsprang, oder nach seinem stillern Charakter gelassen davonging [...] (Goethe 2006, 41)

> I walked across the court to a well-built house, and ascending the flight of steps in front, opened the door, and saw before me the most charming spectacle I had ever witnessed. Six children, from eleven to two years old, were running about the hall, and surrounding a lady of middle height, with a lovely figure, dressed in a robe of simple white, trimmed with pink ribands. She held a brown loaf in her hand, and was cutting slices for the little ones all round in proportion to their age and appetite. She performed her task in a graceful and affectionate manner, each claimant awaiting his turn with outstretched hands, and boisterously shouting his thanks. Some of them ran away at once to enjoy their evening meal, whilst others of a gentler position retired [...]. (Goethe 1854, 260)

Werther is struck by the idyllic scene and enjoys the "Schauspiel" [spectacle, theater play] which, in its innocence, precisely lacks all theatrical quality: "[M] eine ganze Seele ruhte auf der Gestalt, dem Tone, dem Betragen [...]." (Goethe 2006, 41) [My whole soul was absorbed by her air, her voice, her manner [...]. (Goethe 1854, 260).] This scene is probably the most famous of the entire novel. It has regularly been depicted in illustrated editions of the eighteenth and nineteenth centuries. Particularly interesting in the context of the present study is Julius Nisle's 1840 *Werther* series, which focuses on moments of secret and intense voyeurism in Goethe's novel. Nisle shows the scene of Werther drawing the children (Fig. 8) and of Werther watching Lotte (Fig. 9) as the first and second illustration of the novel. He thus implicitly points to a connection between these two scenes. As the sequence of the two illustrations suggests, Werther's love for Lotte is born out of an attitude of spectatorship. It arises as the attempt to become one with the object of his contemplation. Pursuing his love for Lotte, Werther acts on the call of the visual.

In his engraving of the supper scene, Nisle places Werther in the door, on the margins of the scene. He has Lotte turn her back on Werther, and he humorously mirrors Werther's absorption in that of a dog that is in the foreground of the image and that stares at a piece of bread that one of the children eats. Other than this child, only one other child has turned away from Lotte. Equipped with an extravagant hat, a toy horse in his right hand and a sword in his left hand, he seems to be so absorbed in playing that he does not care to take

DIE LEIDEN DES JUNGEN WERTHER.

II.

Figure 9: Julius Nisle, No. 2 of 12 steel engravings to *Die Leiden des jungen Werthers* (Nisle 1840). Source Klassik Stiftung Weimar / Herzogin Anna Amalia Bibliothek / F 3337 (f).

part in the idyllic supper scene. We will see this child – who, while being Nisle's own invention, interestingly reflects central aspects of the novel's organization of the visual – return to occupy an important position in a later illustration in Nisle's series.

It is difficult to ascertain Werther's precise attitude in Nisle's illustration of the supper scene, for it is not clear from the engraving whether Werther is actually watching Lotte and her siblings. His head points slightly away from the group, making it appear just as likely that he is lost in thoughts, already imagining having taken a position within the scene (see Assel 2018).

Werther never quite recovers from the overwhelming effect that the supper scene has on him. Indeed, Werther is subsequently haunted by the desire to enter the idyllic scene, and this desire motivates the sequence of the narrated events. The novel's 'First Supper' attains in this way an almost religious importance for the rest of the novel; and the bread handed out at the first supper waits to be complemented by the bread and wine that constitutes Werther's last meal before his suicide. The first step toward the movement into the scene actually occurs already in the very description of this scene, for when Werther writes about the supper, he is having supper himself. Werther thus superimposes the scene of writing on the described scene:

> Da bin ich wieder, Wilhelm, will mein Butterbrod zu Nacht essen, und dir schreiben. Welch eine Wonne das für meine Seele ist, sie [Lotte] in dem Kreise der lieben muntern Kinder, ihrer acht Geschwister zu sehen! (Goethe 2006, 39)

> I have just returned, Wilhelm, and whilst I am taking supper I will write to you. What a delight it was for my soul to see her in the midst of her dear, beautiful children—eight brothers and sisters! (Goethe 1854, 259)

In drawing attention to his own supper, Werther essentially forces his addressee Wilhelm to picture him, Werther, as having supper and thereby to move him into the center of the supper scene that he actually witnesses only from the outside.

In later scenes at Lotte's house, Werther is allowed to distribute the bread for Lotte's siblings. He thus assumes the central role in the tableau that he had previously contemplated as an outsider: "Ich schnitt ihnen das Abendbrod, das sie nur so gern von mir als von Lotten annehmen [...]." (Goethe 2006, 103) [I waited upon them at tea,[6] and they are now as fully contented with me as with Charlotte [...]. (Goethe 1854, 287)]

Interestingly, Julius Nisle's *Werther* series does not include an engraving of Werther providing supper for the children. In its place, he offers the reader an illustration of Werther playing with the children – a scene whose structure presents a curiously distorted repetition of the original supper scene. At first, the

---

6 In the orginal, "Abendbrot" [supper, literally, evening bread].

parallel between the two scenes seems to hold up well enough: Werther has taken Lotte's position amidst the children, while a visiting doctor assumes the role of the outside spectator:

> Vorgestern kam der Medicus hier aus der Stadt hinaus zum Amtmann, und fand mich auf der Erde unter Lottens Kindern, wie einige auf mir herumkrabbelten, andere mich neckten, und wie ich sie kitzelte und ein großes Geschrey mit ihnen erregte. (Goethe 2006, 59)

> The day before yesterday, the physician came from the town to pay a visit to the Judge. He found me on the floor playing with Charlotte's children. Some of them were scrambling over me, and others romped with me, and as I caught and tickled them they made a great noise. (Goethe 1854, 268)

What disturbs the attempt at a true repetition, however, is the fact Werther is entirely aware of his spectator and that this spectator does not remain standing in simple admiration. In fact, the doctor is much less pleased with Werther's foolish games than Werther was with Lotte's maternal behavior, and Werther notices his discontent. Nonetheless, Werther stubbornly remains in the posture of the undisturbed and caring family member, seemingly oblivious to his critical audience. It is as if he wanted to resemble Lotte at all costs:

> Der Doctor, der eine sehr dogmatische Dratpuppe [sic] ist, unterm Reden seine Manschetten in Falten legt und einen Kräusel ohne Ende herauszupft, fand dies unter der Würde eines gescheuten Menschen; das merkte ich an seiner Nase. Ich ließ mich aber in nichts stören, ließ ihn sehr vernünftige Sachen abhandeln; und baute den Kindern ihre Kartenhäuser wieder, die sie zerschlagen hatten. (Goethe 2006, 59)

> The Doctor is a formal sort of personage; he adjusts the plaits of his ruffles, and continually settles his frill whilst he speaks with you, and he thought my conduct beneath the dignity of a sensible man. I could perceive this by his countenance. But I did not suffer myself to be disturbed. I allowed him to continue his wise conversation whilst I rebuilt the children's card-houses for them as fast as they threw them down. (Goethe 1854, 268)

What turns Nisle's engraving of this scene (Fig. 10) into something like a parody of the supper scene is that Nisle stresses Werther's failure to assume the desired position of Lotte. Again, there is an outside spectator in the doorframe who is looking at a scene of one adult among many children. However, whereas in the earlier scene, the children's attention focuses by and large on the adult *in* the room – i.e. Lotte – not a single child looks at Werther in the later scene. Werther, while being among the children, is essentially marginalized and remains the spectator he was before. Unlike Lotte, who stood turned away from the intruding spectator, Werther directs his gaze, as far as one can tell, either to the stern spectator in the door, or at least to the spectator's caricature within the

DIE LEIDEN DES JUNGEN WERTHER.

IV.

Figure 10: Julius Nisle, No. 4 of 12 steel engravings to *Die Leiden des jungen Werthers* (Nisle 1840). Source Klassik Stiftung Weimar / Herzogin Anna Amalia Bibliothek / F 3337 (f).

room – the boy in the pose of the preaching philistine (see Assel 2018). Dressed in a prominent hat and holding up a stick or sword, this boy reminds us of the playing child from the earlier scene – the only child that cared neither about the bread nor about Lotte. This boy, who, absorbed in his game of playing the knight, was at the margins of the earlier engraving, is now at the center of attention, leaving us with the impression that Nisle meant to suggest some sort of alternative to Werther's failing attempts to assume this central position himself. If immersion in the scene is to be attained, it is not through spectatorship but through the practice of play.

Leaving behind the particular sequence of scenes related to the supper scene and looking at the novel more broadly, it appears that among the factors that contribute to Werther's failure to become immersed in the scenes that he beholds, change and temporality are of the greatest importance. For the possibility of Werther's immersion through seeing relies on a pronounced stability of the image, unchanging over time. Only if the image does not change does Werther have a chance to enter it. The temporal extension of spectatorship is a prominent theme in the sequence of scenes that describes Werther's drawing of the children, who remain motionless for one or two hours. However, something similar is true also for the scene in which Werther sees Lotte among her siblings. Werther's subsequent attempts to assume Lotte's position among the children suggest a remaining validity of the image even in the absence of the image itself. The moment in which Werther realizes that, generally, nature does not allow for such stability is a moment of deep-felt crisis:

> Es hat sich vor meiner Seele wie ein Vorhang weggezogen, und der Schauplatz des unendlichen Lebens verwandelt sich vor mir in den Abgrund des ewigoffenen Grabes. Kannst du sagen: Das *ist!* da alles vorüber geht? da alles mit der Wetterschnelle vorüber rollt, so selten die ganze Kraft seines Daseyns ausdauert, ach! in den Strom fortgerissen, untergetaucht und an Felsen zerschmettert wird; Da ist kein Augenblick, der nicht dich verzehrte und die Deinigen um dich her, kein Augenblick, da du nicht der Zerstörer bist, seyn mußt; der harmloseste Spaziergang kostet tausend armen Würmchen das Leben; es zerrüttet Ein Fußtritt die mühseligen Gebäude der Ameisen, und stampft eine kleine Welt in ein schmähliches Grab. [...] Und so taumle ich beängstigt. Himmel und Erde und ihre webenden Kräfte um mich her: Ich sehe nichts als ein ewig verschlingendes, ewig wiederkäuendes Ungeheuer. (Goethe 2006, 107–109)

> It is as if a curtain had been drawn from before my eyes; and, instead of prospects of eternal life, the abyss of an ever open grave yawned before me. Can we say of anything that it exists when all passes away—when time, with the speed of a storm, carries all things onward—and our transitory existence, hurried along by the torrent, is either swallowed up by the waves or dashed against the rocks. There is not a moment but preys upon you, and upon all around you—not a moment in which you do not yourself become a destroyer. The most innocent walk deprives of life thousands of poor insects; one step destroys the

fabric of the industrious ant, and converts a little world into chaos. [...] [S]o [...] I wander [*taumle*, more literally: stagger] on my way with aching heart, and the universe is to me a fearful monster, forever devouring its own offspring. (Goethe 1854, 289)

Facing the destructiveness of an ever-changing nature, Werther can but "stagger full of fright" [taumle ich beängstigt] from scene to scene. In contrast to Büchner's Lenz, who, as we shall see, strives to affirm and practice the confrontation with an ephemeral nature in which nothing *is* but the constant movement from one image to the next, Werther perceives the realization of perpetual change as a threat to his way of seeing and being. Werther's way of seeing the world and being in the world cannot account for change. Werther relies on the stability of being as the condition of possibility of his progressive entrance into the witnessed scene. At the moment when this stability is lost, the movement *into* the tableau becomes a stagger from one tableau to the next.

This inability to perceive change also disqualifies Werther as an observer: he stops at the description of the static tableau. Failing to observe, Werther, the writer of letters, also fails to become a narrator: a narrator who is terrified by the idea of change would be utterly absurd. Narration relies precisely on the existence of a chain of events.[7] To put it differently, Werther exemplifies a visual regime (of absorption) in which observations (both as acts of seeing and as literary procedures that combine description with narration) are not possible.

Admittedly, there are moments in the novel in which Werther operates almost as a traditional narrator. However, Werther marks these moments explicitly as foreign particles in his letters, motivated as a concession to his addressee, Wilhelm. The code words for this inauthentic mode of writing are "history" and "chronicling":

- [D]er Brief wird dir recht seyn, er ist ganz historisch. (Goethe 2006, 23) [This letter will please you: it is quite a history. (Goethe 1854, 253)]
- Ich bin vergnügt und glücklich und also kein guter Historienschreiber. (Goethe 2006, 37) [I am a happy and contented mortal, and so[8] a poor historian. (Goethe 1854, 259)]
- [S]o hast du hier lieber Herr, eine Erzählung, plan und nett, wie ein Chronikenschreiber das aufzeichnen würde. (Goethe 2006, 141) [I send you, my dear

---

7 Werther's limitation to the perception of images and his inability to integrate these images into a meaningful narrative has been highlighted, in a different context, also by Caroline Wellbery. Wellbery analyzes Werther's isolation of individual images as an important feature that sets Goethe's novel apart from the tradition of the sentimental novel (Wellbery 1986, 237–239).
8 Boylan translates "but" instead of "and so" (*und also*).

sir, a plain and simple narration of the affair, as a mere chronicler of facts would describe it. (Goethe 1854, 303)]

All history writing, all chronicling of events is condemned as being foreign to Werther's own artistic self, which strives toward moments of full immersion in a single, seen image. This contrast to the figure of the chronicler is important, as it clarifies the position of the novelistic subject. Both the chronicler and the Wertherian absorbed spectator are figures operating from the margins of the scene. But whereas the chronicler affirms this position permanently and thus becomes capable of narrating sequences of events, the Wertherian spectator tentatively tries to overcome his distant position. He wants to enter the scene and therefore cannot allow for its passing.

If there is narration in *Die Leiden des jungen Werthers* beyond those passages in which Werther assumes the despised role of the chronicler/historian, it is in the narration of his failing attempts to enter the perceived tableaux, and in Werther's staggering movement from one tableau to the next. Within the text of the letters, we can necessarily witness only Werther's failure, for every successful attempt to enter the scene would coincide with the spectator's self-annihilation. Werther's writing is preoccupied with scenes that Werther witnesses from the outside and has yet to enter. Once in them, there will not only be no narration, but also no spectator and no description any more. The whole novel and every word uttered in it are in this sense an expression of the subject's failure to be one with the world (a world, it bears emphasizing, that is, in Lukács's sense, wholly misconstrued as a static image).[9]

The verb that Werther uses to describe the movement from one tableau to the next is *taumeln* [stagger]: "Und so taumle ich beängstigt." The same word is used again at the end of the novel as an odd metaphor in Werther's proclamation of his suicide: "Hier Lotte! Ich schaudre nicht, den kalten schrecklichen Kelch zu fassen, aus dem ich den *Taumel* des Todes trinken soll!" (Goethe 2006, 263, my emphasis) [See, Charlotte, I do not shudder to take the cold and fatal cup,

---

9 To be sure, there are several points in the novel in which transient moments of successful immersion are suggested. Without exception, they coincide with moments of silence and inactivity. The first such incident can be found at the beginning of the second letter, where Werther announces both his happiness and his lack of artistic productivity: "Ich bin so glücklich, mein Bester, so ganz im Gefühl von ruhigem Daseyn versunken, daß meine Kunst darunter leidet." (Goethe 2006, 15) [I am so happy, my dear friend, so absorbed in the exquisite sense of mere tranquil existence, that I neglect my talents. (Goethe 1854, 249)] On incidents of silence in Goethe's novel, see Müller-Salget 2005, 77 and Dotzler 1999.

from which I shall drink the draught[10] of death. (Goethe 1854, 353)] Admittedly, the meaning of the word *Taumel* differs between the two instances. In the first case, *Taumel* is used as a verb to describe an uncontrolled movement; in the second case, Taumel appears as a noun, and it evokes a feeling of intoxication or dizziness – Grimm's *Deutsches Wörterbuch* [German Dictionary] lists the word "Taumelbecher" for "becher mit berauschendem trank" (Grimm 1854–1961, vol. 21, column 203) [cup filled with intoxicating drink]. Nevertheless, the reoccurrence of the otherwise relatively rare word in two prominent scenes of the novel is remarkable enough to justify the close association of the two scenes. The *Taumel* sets in at the moment when Werther realizes his futile movement from one tableau to the next, and it leads right to Werther's death. The *Taumel* epitomizes Werther's failure to account for change in the tableaux that he contemplates. Unable to account for change, Werther, by implication, falls short of becoming an observer, and his letters, which capture his experience of the world, lack coherent narrative development.

Fittingly, the final part of Goethe's epistolary novel is told by the fictional editor of Werther's letters, and not through Werther's letters directly. This transition to the voice of the editor is necessary not only because Werther cannot possibly give an account of his own death, but also because Werther has, even prior to his death, proven his inability to narrate. His eventual inability to tell the story of his own death only accentuates his prior inability and unwillingness to move beyond the describable tableaux, to account for change, and to produce a coherent narrative.[11]

Strikingly, Werther's inability to see change parallels in important aspects what Goethe deplores in his much later writings on the morphology of plants as the contemporary scientists' unwillingness to appreciate the dynamic character of nature. In his short essay "Die Absicht eingeleitet" [The Purpose Set Forth] from 1817 (over 40 years after the first publication of *Werther*), Goethe laments the fact that German scholars describe organic beings largely in terms of their "*Gestalt*" (structured form), or abstract, static essence:

> Der Deutsche hat für den Komplex des Daseins eines wirklichen Wesens das Wort Gestalt. Er abstrahiert bei diesem Ausdruck von dem Beweglichen, er nimmt an, daß ein Zusammengehöriges festgestellt, abschlossen und in seinem Charakter fixiert sei. (Goethe 1981, 55)

> The Germans have a word for the complex of existence presented by a physical organism: *Gestalt* [structured form]. With this expression they exclude what is changeable and assume

---

10 Boylan's "draught" departs from Goethe's original "Taumel" [literally, stagger].
11 Goethe especially emphasized Werther's failure as a narrator in the revised version of the novel (Flaschka 1987, 190).

that an interrelated whole is identified, defined, and fixed in character. (Goethe 1995, 63, emphasis in the original)

Goethe contends that the attempt to construct the *Gestalt* of any plant blatantly overlooks the dynamic aspect of nature. In a language that curiously echoes Werther's realization of the ever-changing character of nature – albeit now in a decisively more positive and affirmative tone –Goethe writes:

> Betrachten wir aber alle Gestalten, besonders die organischen, so finden wir, daß nirgend ein Bestehendes, nirgend ein Ruhendes, ein Abgeschlossenes vorkommt, sondern daß vielmehr alles in einer steten Bewegung *schwanke*. (Goethe 1981, 55, my emphasis)

> But if we look at all these *Gestalten*, especially the organic ones, we will discover that nothing in them is permanent, nothing is at rest or defined—everything is in a flux of continual motion. (Goethe 1995, 63)

Nature, Goethe emphasizes here, changes perpetually. The plant is for Goethe, as Luke Fischer rightly remarks, crucially a "*temporal* organism" (Fischer 2011, 129, my emphasis), which reveals its essence only in a process of growth and metamorphosis. The plant, Goethe insists, can be understood only if we study it "as a process of formation and not as a static form" (Fischer 2011, 129). Emphasizing the dynamic nature of organic being, Goethe argues for a science that focuses on the notion of *Bildung* (formation, development) instead of *Gestalt* (see Goethe 1981, 55). Like Werther, the scientist Goethe sees that there is in nature "nirgends ein Bestehendes" [nothing permanent].[12]

The precise wording in this passage from "Die Absicht eingeleitet" evokes, however, not only Goethe's first novel, but, as Dorothea Kuhn pointed out long ago, also Goethe's most famous play, *Faust* (Kuhn 1952/1953, 347–348, Goethe 2005, vol. 2, 152). The correspondence between "Die Absicht eingeleitet" and *Faust* is worth fleshing out here. In his morphological essay, Goethe says that in nature everything always "schwanke." Miller's translation – "flux of continual motion" – does not do full justice to the term "schwanke," as it covers over the striking resonance with the famous first line of the "Dedication" to *Faust:* "Ihr naht euch wieder, *schwankende* Gestalten [...]." (Goethe 2005, vol. 1, 11, my emphasis) [Ye hover nigh, dim-floating shapes again [...]. (Goethe 1970, 4)] Al-

---

12 See also Goethe's remarks in his essay "Wolkengestalt nach Howard" [Cloud Formations According to Howard]. Here, Goethe emphasizes that in order to test the validity of Luke Howard's system of cloud classification, he has to see to which extent Howard's system facilitates the temporally extended observation in nature (see also Förster 2012, 276 and Vogl 2005).

though the context in the dedication is a very different one – here the poet is facing the return of fictional characters that had occupied him earlier in his life – the parallel remains important: the challenge that the poet of the dedication of *Faust* faces resembles the challenge of Werther as well as that of the natural scientist: it is the challenge to come to terms with the absence of any fixed state in nature.[13] Seeing nature means seeing change, motion.

However, in Goethe's view, the study of plants does not stop at the realization of the ever-changing nature of organic beings. Rather, this realization has to be complemented by the understanding of an inherent formal unity in this dynamic progress. Goethe's study of plants is characterized by a seeing that perceives development *and* identity. More precisely, the assertion of an underlying identity in the transformation of plants is a necessary presupposition for Goethe's claim that there is *Bildung* (or development) in nature at all – and not merely random change. This duality of change and identity, which is crucial to Goethe's concept of Bildung, can be clearly studied in his essay *Versuch die Metamorphose der Pflanzen zu erklären* [Metamorphosis of Plants]. Here, Goethe observes the growth of the different organs of a plant as the continued variation of one and the same form. From leaves to sepals, petals, and stamens, there is, in Goethe's account, only one recurrent form in variation (see also Bortoft 1996, 77–89).

Goethe's view of Bildung in nature finds its most canonical expression in his peculiar and much-discussed concept of the *Urpflanze* [original plant]. In Goethe's view, any given plant can be constructed out of the general knowledge of this most original plant structure. Goethe's Urpflanze is both a regulative general principle of plant development and in itself a visually perceptible entity: the "archetype or idea of the plant kingdom, which comes to differentiated expressions in the many species of plants" (Fischer 2011, 130). This is the remarkable feature of Goethe's theory of the observation of nature: to the true observer, nature reveals itself visually in both its dynamic and stable image. For Goethe, the stability behind the change is, in other words, not just a construct of intellectual abstraction, but itself open to empirical experience: the Urpflanze itself, according to Goethe, can be seen (Fischer 2011, 130; Bortoft 1996, 261–289). Or, as Eckart Förster argues in his prominent recent reading of Goethe's morphology in his

---

[13] Albrecht Schöne, however, contends that, in contrast to Goethe's morphological writings, in the dedication to *Faust* there prevails the idea that the "schwankende Gestalten" can eventually be fixed in the finished poetic work (Goethe 2005, vol. 2, 152).

book *The Twenty-Five Years of Philosophy*, the underlying universal concept actually first enables any correct individual momentary perception.[14]

This aspect of Goethe's morphology – i.e. the idea that the awareness of change has to be accompanied by a simultaneous awareness of an underlying stable image – has, it should be noted here, no direct parallel in the narrative reflections of Goethe's *Werther*. However, this idea does shape, as I will argue in the following section, a text that is conventionally understood to be influenced by *Werther*, namely Georg Büchner's novella *Lenz*. Unlike Goethe in his *Werther*, Büchner focuses not primarily on the difficulty of giving up on the idea of stable images. Instead, he emphasizes that the realization of change has to be complemented by a simultaneous awareness of unity and stability.

Reading Büchner will not only allow us to study one more narratological meditation on the problem of observation; it will also lead to a stronger synthesis between scientific and literary reflections on observation, which remain only loosely connected in the work of Goethe. For what does it really signify, in the case of Goethe, that a scholar produces ideas in his scientific writings that show some parallels to a fictional narrative published more than 40 years earlier? Recognizing these parallels certainly does not shed much new light either on my reading of *Werther*, nor does it, conversely, really allow us to say that we have identified in *Werther* an origin of Goethe's scientific views (although one might still be tempted to think in that direction). *Werther* and the study of morphology remain distinct. One could, to be sure, assert here that the parallels between Goethe's literary and scientific views justify the present study's underlying thesis that observation (in my sense of the term, as the combination of the seeing of images and of dynamic sequences) is both a scientific and literary concern. Without having to trace clear influences, we can see that the question of how to proceed from the seeing of an image to the seeing of sequences takes a central place in very different discourses of the eighteenth and nineteenth centuries. But in the case of Büchner, this parallel quest for observation in literature and the sciences becomes much more evident. In Büchner, we can study a scholar who works on morphology simultaneous to his literary musings on observation, and whose literary works bear philologically unambiguous traces of Goethe's and his own scientific theories.

---

14 See Förster 2012, 271–276. Förster's argument is maybe more evident in the case of Goethe's theory of colors. As Förster shows, Goethe thought that whenever we see a color, the eye actually also produces the complementary color and thus completes the circle of colors: "It is only in the context of the specific 'whole' to which they belong that the individual angle and the individual color are what they are: the whole makes the individual part possible and determines it [...]." (Förster 2012, 271).

Lenz

The problem of the integration of a multitude of evanescent images is, I argue, at the heart of Georg Büchner's unfinished novella *Lenz*, which recounts the historical episode of the ten-day stay of the mentally unstable *Sturm und Drang* writer Jakcb Michael Reinhold Lenz (1751–1792) in the home of the then famous protestant pastor Johann Friedrich Oberlin (1740–1826) in a village in the Alsace region. Significantly, Büchner, like Goethe, was much interested in the study of morphology, and his research essentially followed Goethe's theory. The fact that Büchner's *Lenz* addresses a problem of the procedure of literary observation complementary to the one that Goethe presents in *Werther* should therefore, I argue, be understood not only in terms of a direct influence of the novel *Werther* on the novella *Lenz* – this influence is well established in the secondary literature – but also as a result of the common morphological interests of the two writers.

In his 1836 "Probevorlesung" [trial lecture] at the University of Zürich, "Über Schädelnerven" [On Cranial Nerves], which summarizes the findings of his dissertation, Büchner evokes Goethe's *Versuch die Metamorphose der Pflanzen zu erklären* as a model of scientific inquiry (Büchner 1999, vol. 2, 160). Büchner's own work participates in the endeavor of Goethean morphology and comparative anatomy to find the common form behind a series of variations. Similarly to Goethe, Büchner strives "nach einer gewissen Einheit, nach einem Zurückführen aller Formen auf den einfachsten primitiven Typus" (Büchner 1999, vol. 2, 160) [toward a certain unity, toward the derivation of all forms from the simplest primitive type (Büchner 2012, 176)]. But whereas Goethe justifies the project of morphology first of all in contradistinction to those theories that refuse to acknowledge the dynamic nature of being (that focus on *Gestalt* instead of on *Bildung*), Büchner sees the main strength of morphological research in its ability to integrate the otherwise incoherent collections of different images of nature. As Büchner points out, the knowledge of primal forms allows one to see "zusammenhängende Strecken" (Büchner 1999, vol. 2, 159) [coherent pieces (Büchner 2012, 176)] where there are otherwise merely "getrennte, weitauseinanderliegende facta" (Büchner 1999, vol. 2, 159) [widely separated facts (Büchner 2012, 176)]. Büchner, in other words, takes the perception of variety and change in nature for granted. What one has to learn to observe is the underlying unity – the "coherent pieces." This knowledge of the underlying unity is crucial to understanding that we witness in nature's change a coherent development.

Büchner's novella *Lenz* works in narrative form through this very point that the perception of change has to be accompanied by the identification of a unifying image. And it is precisely in the attempt to supplement the series of separate visual impressions with an underlying image that the story's protagonist fails. In contrast to Werther, Büchner's protagonist Lenz does not struggle with change as such. He affirms the stagger and the ever-changing image of nature as the object of his perception. Whereas Werther experiences his realization of instability in nature as a moment of crisis, Lenz thinks of nature from the outset as ephemeral.

As I will show in this section, the realization of the ephemeral character of nature and the struggle to find a stable underlying form constitute not only a decisive element in the novella's central *Kunstgespräch* [conversation on art], in which Lenz presents his aesthetic theory, but it also defines Lenz's mental instability, and it motivates the disintegrated, paratactic style in which the novella is written. In my reading, the striking paratactic style is an integral part of the novella's aesthetic and narratological agenda, relating to the question of how to perceive series of images. This style should therefore not be attributed merely to the status of the novella as unfinished. Neither should it be reduced to a remnant of Büchner's main source for the story, Pastor Oberlin's report of Lenz's stay with him. As we shall see, Büchner reinforces – albeit only slightly – the already largely paratactic style of Oberlin's report. More importantly, the paratactic style of Oberlin's report gains programmatic significance only in Büchner's novella – namely through the connection with the theoretical Kunstgespräch, which has no parallel in Oberlin's account.

In their readings of the novella's Kunstgespräch, however, critics sometimes overlook the central struggle for a form of seeing that includes both change and identity. Instead, they stress the figure of the petrifying Medusa, which, as we shall see in a moment, Lenz also mentions in his remarks. Lenz cites the myth of Medusa to describe our relation to incidents of beauty in the everyday. Seeing beauty in nature, Lenz argues, we wish to have the powers of Medusa to petrify it and share it with others. All in all, however, this idea of petrification plays, I argue, only a transitory role in Lenz's aesthetic theory.

Lenz begins his explanations of his ideal vision of art in the Kunstgespräch by describing a recent walk in the mountains. While walking, Lenz saw two girls who were sitting on a stone, arranging each other's hair:

> Wie ich gestern neben am Tal hinaufging, sah ich auf einem Steine zwei Mädchen sitzen, die eine band ihre Haare auf, die andre half ihr; und das goldne Haar hing herab, und ein ernstes bleiches Gesicht, und doch so jung, und die schwarze Tracht, und die andre so sorgsam bemüht. (Büchner 1999, vol. 1, 234)

> As I went by the valley yesterday, I saw two girls sitting on a rock, one was putting up her hair, the other was helping her; and the golden hair hung down, and a serious, pale face,

and yet so young, and the black dress, and the other one working with such care. (Büchner 2012, 91)

Lenz claims that this scene in the mountains is unsurpassed in its beauty even by the paintings of the "altdeutsche Schule" (Büchner 1999, vol. 1, 234) [Old German School (Büchner 2012, 91)]. Watching this scene, Lenz imagines being the head of Medusa in order to retain this momentary beauty and to show it to other people:

> Die schönsten, innigsten Bilder der altdeutschen Schule geben kaum eine Ahnung davon. Man möchte manchmal ein Medusenhaupt sein, um so eine Gruppe in Stein verwandeln zu können, und den Leuten zurufen. (Büchner 1999, vol. 1, 234)
>
> The most beautiful, most intimate paintings of the Old German School barely hint at it. At times one would like to be a Medusa's head in order to transform such a group into stone and summon everyone to see it. (Büchner 2012, 91)

Such petrifying force would be necessary if one wanted to keep this image because nature never allows for a long contemplation of any phenomenon. Only nature that has been violently brought to a halt offers itself to sustained and shared contemplation. Evidently, scenes such as the one in *Die Leiden des jungen Werthers* in which the protagonist is allowed to watch and draw the living *natura morta* of the two boys who are sitting on the ground for hours have no place in *Lenz*.

In his reading of the Kunstgespräch, Robert C. Holub focuses on the Medusa to stress the paradox that realist art has to kill in order to show life: "Capturing life in art unavoidably involves a removal of and from life, because the nature of aesthetic reproduction is representation in lifeless appearances." (Holub 1985, 119) Conditioned by both a long-standing theory of realism that emphasizes the minute description of states as the hallmark of realist writing and by a discourse on photography that revolves around the very paradox of life and death in representation that Holub evokes (see, for instance, Benjamin 1991, Barthes 1981), we will possibly all too easily follow Holub's argument. However, it is important to note that Lenz's explanations of the form of realist art that he envisions do not stop at this moment of petrification. Rather than dwell on – and petrify – the image of the girls arranging their hair, Lenz immediately adds another image: "Sie standen auf, die schöne Gruppe war zerstört; aber wie sie so hinabstiegen, zwischen den Felsen, war es wieder ein anderes Bild." (Büchner 1999, vol. 1, 234) [They stood up, the beautiful group was destroyed; but as they climbed down among the rocks they formed another picture. (Büchner 2012, 91)] A Medusa would not grasp reality's true beauty because she does not focus on reality's essentially ephemeral nature – a nature that produces time and again "wieder ein anderes Bild" [again another picture]. The art that Lenz envisions could not be captured by a Medusa:

> Die schönsten Bilder, die schwellendsten Töne, gruppieren, lösen sich auf. Nur eins bleibt, eine unendliche Schönheit, die aus einer Form in die andre tritt, ewig aufgeblättert, verändert [...]. (Büchner 1999, vol. 1, 234)

> The most beautiful pictures, the richest sounds group together and dissolve. Only one thing remains, an endless beauty moving from one form to another, eternally unfolding, changing [...]. (Büchner 2012, 91)

What becomes clear in this passage, however, is not simply that, in Lenz's account, the reality of nature's true beauty consists in a continuous stream of changing images. Rather, nature (and therefore ideal art) consists in a series of images that is united by the fact that each passing image is an instantiation of one eternal beauty. Failing to capture the importance of the underlying eternal beauty, Robert Holub distinguishes simply between life's constant movement and art's restriction to dead images of single moments in time:

> Eternal change versus a total motionlessness, pulsation versus petrification, life versus death, and God versus the Medusa are the antinomic poles to which Lenz's realism is ineluctably attracted. (Holub 1985, 120)

Holub's opposition of "life versus death" fails to do full justice to Lenz's theory of both art and nature as revolving around the ideal of *one* eternal beauty ("eine unendliche Schönheit") that realizes itself in swiftly changing images. For the Lenz of Büchner's novella, reality consists neither simply in a constant flow, nor in a dead image.

We can recognize in Lenz's concept of reality again the language of the student of morphology. Life, according to Goethe's and Büchner's understanding of morphology, constantly changes from one arrangement to the next – but in all these different shapes one constant form remains recognizable. There always remains one eternal beauty. In Goethe's terminology, this one eternal beauty corresponds to the *Urpflanze*. While the connection between morphology and literary representation remains loose in the case of Goethe, it is far more evident in the case of Büchner. Not only did Büchner work on *Lenz* more or less simultaneously to his morphological research, but his formulation in the Kunstgespräch also echoes very clearly the morphological idea of the existence of one underlying structure that defines all organisms in all of their stages of development. In insisting on the influence of morphology, I am not denying, of course, that Büchner draws in *Lenz* also on a range of other discourses: art criticism as well as the medico-psychological discourse, which is evident in the portrayal of Lenz's mental instability, certainly come to mind here. And I will have more to say about the connection between psychology and the literary device of observation below. But let us here concentrate first on the paradigm of morphology. What is most striking in this regard is the difference between the programmatic Kunstgespräch and Lenz's behavior in the rest of the novel. Lenz fails to fulfill the kind of seeing that he prescribes in the Kunstgespräch. While Lenz expresses in the Kunstgespräch a concept of art and vision that strikingly resembles Goethe's and Büchner's own scientific theories of morphology, Lenz's actual perceptions fail to implement this way of seeing, which integrates the perception of identity and of change. Lenz fails to perceive the common form behind eternal change. He sees only a series of disconnected images.

Lenz's failure to fully implement his ideal of observing both change and identity is reflected in the stylistic structure of Büchner's novella. The overwhelmingly paratactic syntax and the scarcity of conjunctions in the novella, which is well captured in Henry J. Schmidt's English translation, is a product of Lenz's disconnected perceptions. The sentences in the novella consist of long chains of short phrases of equal value. In some cases, it seems that sheer physical exhaustion, not grammatical structure or conceptual unity, determines the placement of a period.

To avoid a misunderstanding, it should be recalled that in *Lenz* the story is presented by a heterodiegetic narrator. In other words, we do not hear Lenz speak directly (as is the case in *Werther*). However, the story is strongly internally focalized and thus tracks Lenz's thoughts and perceptions in a way that allows us to analyze the stylistic features of the text as a (more or less) direct expression of the perceptions of the character Lenz.

Consider, for instance, the four sentences relating the events on the evening of Lenz's arrival at Oberlin's parsonage. After dinner, Oberlin assigns Lenz a room in the schoolhouse: "Endlich war es Zeit zum Gehen, man führte ihn über die Straße, das Pfarrhaus war zu eng, man gab ihm ein Zimmer im Schulhause." (Büchner 1999, vol. 1, 227) [At last it was time to leave, he was led across the street, the parsonage was too small, he was given a room in the schoolhouse. (Büchner 2012, 86)] Already in this first sentence of this passage, the paratactic structure is prominent. Particularly striking is the fact that the causal structure of the events is almost completely obscured. The apparent reason for Lenz's move from the parsonage to the school – namely that the parsonage is too small – is not marked by any conjunction. Instead, the independent main clause that names the reason is awkwardly inserted in between the two main clauses that describe the movement from the parsonage to the school. In this peculiar construction, Büchner deviates from Oberlin's report, which he follows otherwise relatively closely in this passage. Oberlin merely states: "Wir logirten ihn in das Besuchzimmer im Schulhause." (Büchner 1999, vol. 1, 966) [We lodged him in the guestroom in the school.] Büchner, in other words, extends the account of this scene to add an explanation as to why Lenz was lodged in the school; at the same time, however, he avoids syntactically integrating this causal connection by any conjunction.

The following sentences in Büchner's novella continue and amplify this paratactic structure. In the first of these sentences, only a few appositions and short subordinate clauses interrupt the chain of some twenty main clauses:

> Er ging hinauf, es war kalt oben, eine weite Stube, leer, ein hohes Bett im Hintergrund, er stellte das Licht auf den Tisch, und ging auf und ab, er besann sich wieder auf den Tag, wie er hergekommen, wo er war, das Zimmer im Pfarrhause mit seinen Lichtern und lieben Gesichtern, es war ihm wie ein Schatten, ein Traum, und es wurde ihm leer, wieder wie auf dem Berg, aber er konnte es mit nichts mehr ausfüllen, das Licht war erloschen, die Finsternis verschlang Alles; eine unnennbare Angst erfaßte ihn, er sprang auf, er lief durchs Zimmer, die Treppe hinunter, vor's Haus; aber umsonst, Alles finster, nichts, er war sich selbst ein Traum, einzelne Gedanken huschten auf, es war ihm als müsse er immer „Vater Unser" sagen; er konnte sich nicht mehr finden, ein dunkler Instinkt trieb ihn, sich zu retten, er stieß an die Steine, er riß sich mit den Nägeln, der Schmerz fing an, ihm das Bewußtsein wiederzugeben, er stürzte sich in den Brunnstein, aber das Wasser war nicht tief, er patschte darin. (Büchner 1999, vol. 1, 227–228)

> He went upstairs, it was cold up there, a large room, empty, a high bed in the background, he placed the lamp on the table and walked up and down, he recalled the day just past, how he had come here, where he was, the room in the parsonage with its lights and dear faces, it was like a shadow to him, a dream, and he felt again like on the mountain, but he could no longer fill the void with anything, the light was out, darkness swallowed everything; an unnameable fear seized him, he jumped up, he ran through the room, down the stairs, in front of the house; but in vain, all was dark, nothing, he felt himself to be a dream, isolated thoughts flitted by, he held them fast, he felt he had to keep saying "Our Father"; he could no longer find himself, he beat against the stones, he tore at himself with his fingernails, the pain began to restore him to consciousness, he threw himself into the basin of the fountain, but the water was not deep, he splashed around in it. (Büchner 2012, 86)

The same paratactic structure is taken up again in the following two sentences. First, there is one shorter sentence consisting of three main clauses: "Da kamen Leute, man hatte es gehört, man rief ihm zu." (Büchner 1999, vol. 1, 228) [Then people came, they had heard this, they called out to him. (Büchner 2012, 86)] The last sentence of the passage, finally, ends only at the moment when Lenz, fully exhausted, has calmed down:

Cberlin kam gelaufen; Lenz war wieder zu sich gekommen, das ganze Bewußtsein seiner Lage, es war ihm wieder leicht, jetzt schämte er sich und war betrübt, daß er den guten Leuten Angst gemacht, er sagte ihnen, daß er gewohnt sei kalt zu baden, und ging wieder hinauf; die Erschöpfung ließ ihn endlich ruhen. (Büchner 1999, vol. 1, 228)

Oberlin came running; Lenz had come to his senses, fully aware of his situation, he was at ease again, now he was ashamed and sorry to have frightened these good people, he told them he was used to taking cold baths, and went back up; exhaustion finally allowed him to rest. (Büchner 2012, 86–87)

The paratactic style of the novella, I argue, is a consequence and expression of Lenz's failure to observe: he sees only change; he does not recognize any aspects of stability in this change, which alone would allow him to construct a process of coherent development.

To be sure, in some sense, Lenz's relentless recording of new images approximates a language of factual, realistic documentation. Paradoxically, however, this same act of listing also creates, as the novella proceeds, the opposite effect on the reader. It evokes Lenz's insanity, his inability to integrate his perceptions into a unified narrative. This paradox of a stylistic device that appears to express utmost objectivity while at the same time approximating a state of insanity is maybe best, albeit possibly involuntarily, expressed in Wilhelm Mayer's 1921 article on *Lenz*. Mayer, a trained psychiatrist, first praises Büchner's novella as the "phrasenloseste Dichtung vom Irrsinn, die sich vorstellen lässt" (Mayer 1921, 889) [the least platitudinous literary work on insanity one could imagine] and compares the novella's style even to "ärztliche Journalnotizen" (Mayer 1921, 889) [medical notes]. Hardly a page later, however, Mayer then claims that Büchner's prose skillfully mirrors not the short-hand of the trained doctor, but, instead, the dissociated thoughts of the schizophrenic patient. Mayer's description of Lenz's psychotic mind – a mind that is characterized by abrupt unconnected thoughts – closely resembles that of his prior description of the praised "ärztliche Journalnotizen." In other words, the gesture toward a documentary realism and the mimicry of an insane mind are almost indistinguishable:

Wie er [Lenz] dann aus allen Qualen heraus abrupter, merkwürdiger in seinen Aeusserungen wird, und schliesslich das uns bekannte Bild schizophrener Dissoziation bietet, ist vom Dichter mit einer so unerhörten Einfühlungsfähigkeit geschrieben, dass ich jedem, der sich für das Problem der verständlichen Zusammenhänge innerhalb der schizophrenen Erkrankung interessiert, raten möchte, dieses meisterhafte Stück zu lesen, laut zu lesen, weil es dann noch besser in dem atemlosen Sichfolgen der Sätze und Worte unheimlich in den Zusammensturz eines Menschen sehen lässt. (Mayer 1921, 890)

How he [Lenz] then becomes, due to his suffering, increasingly abrupt, strange in his statements, and how we are finally presented with the image of schizophrenic dissociation that is familiar to us, is written by the poet with an empathy so unheard of that I would like to advise anyone interested in the understandable aspects of schizophrenia to read this masterful piece, to read it aloud, because one can then view in the breathless sequence of sentences and words even better the uncanny collapse of a human being.

The paratactic style of Büchner's novella is important because it functions as a synthesis of several important aspects of this text. This style is a means to express Lenz's insanity, but it also points back to the novella's central Kunstgespräch. Critics so far have been well aware of the connection between the paratactic style and Lenz's insanity, but they have, to my knowledge, not yet described this style also as a (merely partial) implementation of the artistic realism that Lenz envisions in the Kunstgespräch.[15] The paratactic style is a reflection of the way in which Lenz *sees* the world; it is a representation of the world's visuality as a series of quickly changing images. Lenz's vision also approximates observation to the extent that it shows the series of consecutive images as the irreducible object of realist perception and representation. However, Lenz does not arrive at observation proper because he fails to integrate the series of images into coherent and narratable sequences. Narration based on observation proceeds from the initial perception of a static image to the sustained study of the development of this image. Through its sustained focus on an image, observation reveals how the image changes and develops over time. Observation proceeds, as I discussed in my reading of Rétif's *Les Nuits de Paris*, as the consecutive discerning of an image and filming of a coherent sequence. Whereas Werther, as I showed in the previous section, fails to observe because of his unwillingness to account for change in the image, Lenz fails to observe because of his inability to integrate the series of dissociated images that he perceives into coherent narratable sequences. Unable to accompany the perception of a series of images with a simultaneous awareness of a stable underlying image, Lenz also falls behind his own principle of art, which he develops in the Kunstgespräch.

---

15 In different ways, the connections between the Kunstgespräch and the novella as a whole have been repeatedly discussed. For Albrecht Schöne, both Lenz's insanity and his artistic genius (as expressed in the Kunstgespräch) are expressions of Lenz's extreme sensitivity (Schöne 1952, 46). According to John J. Parker, Lenz's vote for realism in the Kunstgespräch is also apparent in his concern for the real suffering of the people around him in the mountains, which is shown in other parts of the novella (Parker 1968, 108). However, other critics, including Peter K. Jansen, insist that there is no direct connection between Lenz's mental disease and his art. Jansen claims that the Lenz of the novella has largely lost his ability to produce art because of his insanity (Jansen 1975, 145). For a review of important interpretations of the Kunstgespräch, see Jansen 1975, 145–147, Holub 1985, 104–105, and, most recently, Borgards 2009, 62–65.

Formulated so abstractly, these cases of failing observations may seem of limited interest. They are, one might say, a formalist curiosity, at best. However, it is important to consider these various failures of observation as an effect of historically important discourse formations. Moreover, if literary observation serves, as I argue, as an important reality effect, producing a world that appears real both in its visuality and in its dynamic development, then it is striking to note that in the context of a range of discourses this reality effect remains impossible (and that, as a consequence, reality cannot be fully captured). In the case of Werther, we see observation fail as a product of what Michael Fried has analyzed in his studies in art history as "absorption." In the case of Büchner, we see how a model of observation shaped by the discourse of morphology is brought in contact with an emerging understanding of psychology: Büchner stages the failure to observe in the way that the morphologist wants us to as a form of insanity. In the final section of this chapter, the problem of observation occurs in yet another cultural context. What I will discuss, more precisely, is the question of why observation (and thus the reproduction of reality, in some sense) was impossible in the newly emerging public spheres of the nineteenth-century metropolis.

"The Man of the Crowd"

Rétif's novel *Les Nuits de Paris* offers an ideal model of how observation combines the description of a momentary image with the narration of an ongoing sequence of events to capture reality. Observation in Rétif's novel is that activity in which a striking image that has been noticed by the still unfocused gaze of the spectator is watched over time so that a sequence of events is disclosed. The nocturnal spectator's observations show us both how a new description can interrupt the narrative and how this description can, in turn, be extended into the narrative. As we have started to see in the two preceding sections of this chapter, the possible pitfalls of this ideal procedure of observation are the topic of many narrative texts in the decades before and after the publication of *Les Nuits de Paris*.[16] Edgar Allan Poe's famous short story "The Man of the Crowd," first published in 1840, half a century after Rétif's novel, is another case in point.[17] "The Man of the Crowd" tells the story of an urban spectator who follows a strange-looking man through the streets of London for twenty-four hours. However, the

---

16 Again, this is not to say that Rétif's novel was a direct reference point for many later writers.
17 "The Man of the Crowd" first appeared in 1840 in *Burton's Gentleman's Magazine*. Poe later revised the story for its publication in the collection *Tales* (1845). My quotations follow the version from *Tales*.

spectator (who is, at the same time, the story's homodiegetic narrator) fails to gain any knowledge about the man, who always remains immersed in groups of strangers around him. The urban spectator can never fully isolate the stranger from the crowd in which he first appeared; he can never fully focus on the man whom he follows and, therefore, he also remains unable to see any meaningful sequence of events. Like Goethe's *Die Leiden des jungen Werthers* and Büchner's *Lenz*, "The Man of the Crowd" thus questions the uncomplicated possibility of the transition from the unfocused openness to be struck by a curious, describable image to the sustained and focused recording of a sequence – and yet the problem that Poe emphasizes is a very different one. *Werther* and *Lenz* highlight the difficulties in the *temporal* shift that occurs in the transition from the seeing of an image to the recording of a sequence. Werther focuses on a single object or scene, as the observer does, but he fails to account for the change that the scene undergoes as time passes. Lenz, by contrast, understands the fleeting nature of any given scene, but he fails to integrate the series of images that he perceives in a coherent narrative sequence. In contrast to Goethe's novel and Büchner's novella, Poe's short story does not locate the difficulty of observation in this temporal shift. The failure of Poe's urban spectator is not a failure to account for change and coherent development; instead, the urban spectator already fails in the prior focusing of the gaze that would allow for the recording of an extended sequence. Poe's observer cannot focus on any given individual person or object. This inability to focus on any one object, to regard it for itself, and to disclose its narrative development, is, I argue, an effect of the particular conditions of urban spectatorship.

Indeed, Poe's short story reflects, in my view, central features of the form of urban spectatorship that emerged in the growing European cities since the late seventeenth century. This kind of urban spectatorship importantly served the orientation among strangers and functioned through a system of classification. The quickly growing cities necessitated such a knowledge of classification because they were in a historically unprecedented way "human settlement[s] in which strangers are likely to meet" (Sennett 1977, 39; see also Fleming 2008, 238). Living among strangers, the inhabitants of the new urban environments had to become experts of classification. They had to learn to gauge the strangers who surrounded them. As John F. Kasson has shown, especially the nineteenth century developed a rich literature of handbooks that were meant to help inhabitants of the metropolitan areas to determine at first glance class and character of their fellow citizens (see Kasson 1990).

Poe's story highlights the epistemic limits of this newly emerging urban knowledge of classification. As Poe's story makes clear, it remains impossible for the urban spectator to focus on any one given person, because the urban spectator categorically does not regard individuals as such; instead, he/she constructs classes and categories, and he/she learns to quickly assign individuals their place in this system of classification. This is the predicament of observation in the city to which Poe alludes already in the story's epigraph by Jean de La Bruyère, the great analyst of the growing Paris of the seventeenth century: "Ce grand malheur de ne pouvoir être seul" [the great misfortune that one cannot be by oneself].[18] I read this epigraph less as an expression of our inability to live alone (although this meaning is clearly present as well), than as a statement about the categorical impossibility of isolating any person in the city and looking at him or her in a way that disregards the presence of the class as whose member he or she appears to the urban spectator in the first place. To be perfectly clear, the problem of urban observation is not so much that one can never isolate a single individual from the physically present crowd; instead, the difficulty consists in categorically seeing anyone by and for himself/herself.[19]

Admittedly, at first sight Poe's story reads very much like another uncomplicated application of Rétif's dyad of the seeing of image and of sequence. "The Man of the Crowd" tells the story of a man sitting in the evening in a coffeehouse in central London and watching the crowd pass by outside in a busy street. After some time, the spectator is suddenly struck by the unique features of an elderly man. He subsequently runs after this man through the streets of London and continues to follow him over the next twenty-four hours. The story itself consists in the retrospective first person account of this pursuit through the English capital.

Summarized in this way, Poe's story seems to be divided into precisely those stages that we learned to tell apart in the reading of Les Nuits de Paris. The initial open-ended curiosity renders the striking, describable, momentary image that is subsequently followed up in the ongoing recording of a sequence of events.

---

18 As Genevieve Amaral has shown, the epigraph comes from the section "De l'Homme" [Of Man] from Jean de La Bruyère's Les Caractères [The Characters] and runs in the original: "Tout notre mal vient de ne pouvoir être seuls: de là le jeu, le luxe, la dissipation, le vin, les femmes, l'ignorance, la médisance, l'envie, l'oubli de soi-même et de Dieu." (La Bruyère 1962, 329, Amaral 2011, 236) [All our misery comes from our inability to be alone: here originates gambling, luxury, dissipation, wine, women, ignorance, slander, envy, oblivion of oneself and of God.]

19 This concern with the problematic attempt to arrive at individualizing observations evokes Michel Foucault's analysis of those new nineteenth-century juridical, medical, and pedagogical techniques that served precisely to develop the knowledge of individual subjects (see Foucault 1977, 251–252, and, more broadly, Foucault 1978; on literature's response to the social and political construction of subjects in the nineteenth century, see Pethes 2012). See also chapter 5 in this book.

However, this pairing of openness and focus, image and sequence is complicated in Poe's story by the fact that the chase through London fails to provide any additional information – let alone a coherent narrative – beyond the initial brief description of the man's strange appearance. In the end, the narrator stops his attempted observation, realizing that his pursuits will produce no results: "He refuses to be alone. *He is the man of the crowd*. It will be in vain to follow; for I shall learn no more of him, nor of his deeds" (Poe 1965, vol. 4, 145, emphasis in the original). To understand what went wrong in the attempt to truly observe the striking individual, we have to understand the peculiar conditions of spectatorship in the modern urban environments.

Like half a century later in Doyle's Sherlock Holmes stories, it is London that serves in Poe's story as the paradigm of the new metropolitan situation.[20] As is well known, in the eighteenth and nineteenth centuries London grew to be the largest city of the world. In some sense, London really became *the* big city in the Western imagination, taking this title from Paris, which had been the Western world's largest city around 1700.[21] For the American writer Poe and his audience – "The Man of the Crowd" was first published in the Philadelphia-based *Burton's Gentleman's Magazine* – London was synonymous with the idea of the modern metropolis. With the choice of London, Poe made his story a story about the modern urban environment as such.

Localizing the central epistemological and narratological crux of the story in the conditions of urban spectatorship allows us to move beyond established interpretations of "The Man of the Crowd." In the past decades, Poe's story has been understood predominately as a text about problems of writing and reading, or, even more generally, as a text about the limits of human reason. Genevieve Amaral, for instance, describes the "The Man of the Crowd" as an allegory of hermeneutic impasses and as "a text resistant to interpretation" (Amaral 2011, 227).[22] In this type of analysis, the narrator's failure to know the man of the

---

20 See, in particular, the beginning of the Sherlock Holmes story "The Adventure of the Blue Carbuncle" (Doyle 1953, 245).

21 Poe's epigraph by the French writer Jean de La Bruyère, whose *Caractères* (1688) arguably presents the first account of spectatorship in the big city, still pays tribute to the Western world's former capital, Paris. I emphasize 'Western' here, as the World's largest city of the seventeenth and eighteenth century was probably Beijing (London outgrew Beijing only in the first half of the nineteenth century).

22 Amaral focuses on the many references to texts and the practice of reading in Poe's story, and she interprets the pursuer's lack of success as a statement about literary communication. Just as the man of the crowd can neither separate himself *from* the crowd nor communicate *with* the crowd, Poe's story has a problematic position in a complex web of intertextuality: the story can neither be said to be completely original nor easily be connected to existing

crowd is understood to be analogous to the reader's difficulty to decipher Poe's text – and indeed any text. A variation of Amaral's position can be found in an essay by J. Gerald Kennedy, who reads Poe's story as a general study of the impasses of human reason:

> Regardless of what we may infer from his actions, the man of the crowd retains the ultimate inscrutability of Melville's white whale, symbolizing (if anything) man's inability to ascertain, by means of reason, any absolute knowledge of the world beyond the self. (Kennedy 1975, 190; see also Sweeney 2003)

While I do not doubt the legitimacy of these 'allegorical' readings of Poe's story, I propose to reconsider "The Man of the Crowd" by taking seriously the narrator's problem for what it is at the surface – a problem of individualizing observation in the big city.

A city, as Richard Sennett laconically defines it in his study *The Fall of Public Man* (1977), is a place "where strangers are likely to meet" (Sennett 1977, 39). This urban coexistence of strangers necessitates a form of mutual gauging and surveillance as a means of orientation. Perhaps the first famous description of urban mutual surveillance can be found in the source for Poe's epigraph, i.e. Jean de La Bruyère's book *Les Caractères* (1688). La Bruyère describes how the citizens of the French capital go every night "aux Tuileries, pour se regarder au visage et se désapprouver les uns les autres" (La Bruyère 1962, 206) [to the Tuileries, to look each other in the face and to disapprove of each other]. This system of mutual judgment is the way in which the citizens sustain their identity and orient themselves among strangers:

> L'on s'attend au passage réciproquement dans une promenade publique; l'on y passe en revue l'un devant l'autre, carrosse, chevaux, livrées, armoiries, rien n'échappe aux yeux, tout est curieusement ou malignement observé ; et selon le plus ou le moins de l'équipage, ou l'on respecte les personnes, ou on les dédaigne. (La Bruyère 1962, 206)

> One watches each other as one passes on a public ride. One undergoes each other's evaluation: carriage, horses, uniforms, coat of arms – nothing escapes the eyes, all is curiously or maliciously observed, and depending on the splendor or meagerness of the equipage one respects the people or disdains them.

In the nightly encounters of strangers in the parks of Paris, everything is surveyed in order to be classified and judged. With Rétif's *Les Nuits de Paris*, I discussed a text that is based precisely on the premises of urban surveillance that

---

texts (Amaral 2011, 234–235; on the question of reading in "The Man of the Crowd," see also Rachman 1997, 656–661).

Jean de La Bruyère lays out. The nocturnal spectator's surveillance of Paris relies on the simultaneous possibility and necessity to survey the anonymous urban crowd. However, Rétif's nocturnal spectator also consistently moves beyond mere classification of strangers. As the nocturnal spectator observes the individuals over time to retrieve their story, he moves from classification to individualized observation and narrative. Poe's "The Man of Crowd," by contrast, forcefully raises the question of whether the form of spectatorship that is characteristic of the modern city truly allows for individualizing observation.

The possibility to zoom in on an individual object, to focus on it and to regard it for itself, which Poe's story ultimately questions, is essential to the transition from the unfocused gaze of the spectator to the sustained and focused recording of a sequence. While the observer must be initially open to being struck by any one person (or object) that strikes his/her attention, he/she also has to learn to focus on a curious person (or object) and to watch it over time. This is the first lesson that the observer has to learn, as Rétif's *Les Nuits des Paris*, and more explicitly still, E.T.A. Hoffmann's last story "Des Vetters Eckfenster" (1822) teach us.

In Hoffmann's story, a man visits his invalid cousin, who spends his days observing from his window the activities on a central marketplace. As the sick cousin invites his visitor to give an account of what he sees, the visiting cousin initially discerns only the mass as such – as an aggregate of colored spots, which he compares to a "vom Winde bewegten, hin und her wogenden Tulpenbee[t]" (Hoffmann 1985 – 2004, vol. 6, 471) [a large bed of Tulips, being blown hither and thither by the wind (Hoffmann 2000, 379)]. Only subsequently does the invalid cousin teach his visitor to fix his gaze on a single object and to follow one object over time: "das Fixieren des Blicks erzeugt das deutliche Schauen" (Hoffmann 1985 – 2004, vol. 6, 472) [you have to focus your attention if you are to see distinctly (Hoffmann 2000, 381)]. Pointing to a curiously clad woman on the marketplace below their window, the older cousin assigns the younger cousin the task of keeping the woman in view as she moves through the crowd:

> Versuche, Vetter, ob du ihren Lauf in den verschiedensten Krümmungen verfolgen kannst, ohne sie aus den Augen zu verlieren; das gelbe Tuch leuchtet dir vor. (Hoffmann 1985 – 2004, vol. 6, 472)

> Cousin, see if you can follow the various twists and turns of her course without losing sight of her. Her yellow head-cloth will be your guide. (Hoffmann 2000, 380)

While Rétif's nocturnal spectator and eventually also Hoffmann's visitor succeed in discerning and observing a single object over time, Poe's narrator in "The Man of the Crowd" fails. He can never fully and categorically divorce "the man of the crowd" from the crowd in which the latter first appeared.

Poe carefully stages in his story the problematic and, in the end, unsuccessful process in which a spectator in the modern city tries to move from the construction of different classes to the focused study of an individual. Poe emphasizes the ultimate failure of the urban spectator to see an individual categorically divorced from the larger group as whose member this individual becomes visible in the first place. This failed transition from class to individual is also a failed transition from unfocused spectatorship to the focused recording that is a necessary part of observation. It is the failure to pass from "a calm but inquisitive interest in every thing" (Poe 1965, vol. 4, 135), as the story's narrator calls it, to the fulfillment of the "the craving desire to keep the man [the man of the crowd] in view—to know more of him" (Poe 1965, vol. 4, 140).

At the very outset of the story, however, Poe's narrator does not even possess the knowledge of classification. He has recovered only recently from a long disease and now encounters society wholly anew. As Charles Baudelaire rightly emphasized in his reading of Poe's story, the freshly cured narrator comes to the world like a newborn baby; he has to learn all over again how to see the world (Baudelaire 2009). Sitting behind a bow window in a coffeehouse in central London and watching the people pass by outside, the narrator initially sees only "masses" (Poe 1965, vol. 4, 135) and thinks of them merely "in their aggregate relations" (Poe 1965, vol. 4, 135). His initial perception is reminiscent of the first impression of Hoffmann's visitor, who saw on the marketplace not individual people, but only a floating "bed of tulips." We may recall here also that Walter Benjamin speculated in his Baudelaire essay that, historically, mankind first had to learn how to perceive crowds. Not unlike some impressionist paintings, Hoffmann's "Des Vetters Eckfensters" and Poe's "The Man of the Crowd" are testimonies to the difficulty of seeing crowds as collectives of distinct people instead of as blurred aggregates (Benjamin 1961, 206).

As time goes by, however, Poe's narrator does gradually learn to discern more details of the people in the thoroughfare outside the coffeehouse. He learns to divide the stream of strangers into a growing number of classes and subclasses. At first, the narrator divides the bypassing crowd merely into two large divisions. There are, according to him, those who are focused and steadily follow their way, and those who are lost in their thoughts. The latter group is especially interesting because their feeling of solitude is, paradoxically, a direct product of their embeddedness in the crowd around them. As the narrator tells us, these people were "restless in their movements, had flushed faces, and talked and gesticulated to themselves, as if feeling in solitude on account of the very denseness of the company around" (Poe 1965, vol. 4, 136). The appearance of this class of people, who believe to be alone precisely because so many strangers surround them, is of crucial importance to the story in general because it already gestures to the later failure of the narrator to become an observer of individuals. As the narrator explains, the members of this class react with great bewilderment when someone happens to run into them: "If jostled, they bowed profusely to the jostlers, and appeared overwhelmed with confusion." (Poe 1965, vol. 4, 136) The members of this group are incapable of coming to terms with the strange fact that their solitude is due to those masses that have now jostled them. These loners in the crowd owe all their individuation to the surrounding crowd and therefore cannot be separated from it. They can never be fundamentally by themselves because they only appear to be by themselves when they are surrounded by others. In a similar way, the narrator will remain unable to see the man of the crowd apart from the crowd in which he first appeared.

After this initial distinction between two main classes in the crowd – those who push their way through the crowd with great determination and those who believe themselves alone on account of the crowd – the narrator enters into a catalogue of more detailed and socially determined subdivisions. He notices clerks (and here he distinguishes the upper clerks from the lower clerks) and pickpockets, gamblers, and "gentlemen who live by their wit" (Poe 1965, vol. 4, 137). There are also poor people, and among these "Jew peddlers" (Poe 1965, vol. 4, 138), "sturdy professional street beggars" (Poe 1965, vol. 4, 138) and "feeble and ghostly invalids" (Poe 1965, vol. 4, 138). There are "women of the town" (Poe 1965, vol. 4, 138), "drunkards" (Poe 1965, vol. 4, 138), and "pie-men, porters, coal heavers, sweeps, organ-grinders, monkey-exhibitors, and ballad mongers" (Poe 1965, vol. 4, 139).

Poe's list provides us with a vivid image of the new social space of the great city, in which people from almost all walks of society are densely packed together. But as detailed as this list may be and as nuanced as the descriptions are on which the classification in this list depends, what the story shows us remains a list of classes or categories, not of individuals. Let us, nevertheless, appreciate this attention to detail for a moment. It has been duly noted in the scholarship that Poe's narrator manages to tell the different professional groups apart by referring to often quite minute physical markers. The gamblers, for instance, are recognizable by "a certain sodden swarthiness of complexion, a filmy dimness of eye, and pallor and compression of lip" (Poe 1965, vol. 4, 137). The upper clerks can be identified by their "slightly bald heads, from which the right ears, long used to pen-holding, had an odd habit of standing off on end" (Poe 1965, vol. 4, 137). This analysis of detail is certainly remarkable and historically new. Although there is a prehistory of such detailed descriptions "in the tradition of physiognomics stimulated by Johann Kaspar Lavater, as well as the literature of *mœurs* of such writers as Jean de La Bruyère" (Kasson 1990, 83), it is important to note that Poe, in contrast to Lavater and La Bruyère, identifies social types, not character traits. Poe's description and classification of social types sets a new standard for the literature of the second half of the nineteenth century. It is taken up most prominently in Doyle's Sherlock Holmes stories, in which the detective's ability to determine the professions of strangers in the streets of London is a recurrent theme (see Kennedy 1975, 187; Sweeney 2003, 5). But it also recurs in the realist novel more broadly – just recall, once more, the detailed description of the schoolboy Charles Bovary at the beginning of *Madame Bovary.*

Poe and Doyle react with their descriptions and classifications to a new social need: with the changing systems of production and trade throughout the eighteenth century, many new professions appeared that were no longer easily recognizable by specific forms of dress or uniform (see Kasson 1990, 70; see, for the sake of contrast, e. g., Le Sage 1973, 99 – 100). Detective fiction by Poe and Doyle responds to the new urge to produce a legible crowd of strangers.[23] However, the objects of these new descriptions remain, by and large, classes and categories, not individuals. In Poe's story "The Man of the Crowd," a *clerk* passes by the window – not a man who may work as a clerk. The surveillance of strangers in the city is a surveillance of classes.

Admittedly, as time passes and night falls over the city, the narrator finally tries to move from the perception of classes to the perception of individuals and thus beyond the limits of urban classification. However, this transition remains extremely precarious and ultimately fails. The narrator even goes so far as to suggest that the transition to the perception of individuals is merely an effect of the street lamps that "thr[o]w over everything a fitful and garish lustre" (Poe 1965, vol. 4, 139) and that "enchained me [the narrator] to an examination of individual faces" (Poe 1965, vol. 4, 139). Like the strobe lights in a modern nightclub, Poe's streetlamps flicker and thus break up the stream of the bypassing crowd into flashes of individual faces. None of the faces, however, persist. They come and go with every 'fit' of the lamps. The same medium that allows for the individual faces to appear, also makes their persistence – and, by implication, their continued observation – impossible.

As brief as the narrator's sighting of the indivuidual passers-by is, he still claims that "in my then peculiar mental state, I could frequently read, even in that brief interval of a glance, the history of long years" (Poe 1965, vol. 4, 139).

---

23 See also Walter Benjamin's discussion of detective fiction as a means of identity construction (Benjamin 1961, 178).

This claim, however, is never substantiated or explained. The task will fall to Doyle's Sherlock Holmes to actually reconstruct the "history of long years" from momentary states of being (see chapter 5 in this book).

It is in the context of this attempt to move from the seeing of classes to the seeing of individuals that the narrator first notices the person whom he later calls "the man of the crowd":

> With my brow to the glass, I was thus occupied in scrutinizing the mob, when suddenly there came into view a countenance (that of a decrepid [sic] old man, some sixty-five or seventy years of age,) – a countenance which at once arrested and absorbed my whole attention, on account of the absolute idiosyncrasy of its expression. Any thing even remotely resembling that expression I had never seen before. (Poe 1965, vol. 4, 139–140)

The appearance of the man of the crowd is different from that of all the other men and women whom the narrator saw before. All classification must fail in face of the "absolute idiosyncrasy" of the man's expression. The only direct description of the man's physiognomy that is initially provided – "(that of a decrepid [sic] old man, some sixty-five or seventy years of age)" – is literally put into parentheses, as if to suggest the inadequateness of its categories to describe the man. In addition to this parenthetical description, the reader is merely given the indirect association of the man of the crowd with the devils in the drawings of the contemporary German illustrator Moritz Retzsch (1779–1857):

> Anything even remotely resembling that expression I had never seen before. I well remember that my first thought, upon beholding it, was that Retzsch, had he viewed it, would have greatly preferred it to his own pictural [sic] incarnations of the fiend. (Poe 1965, vol. 4, 140).

Retzsch was internationally best known for the drawings that he designed for Goethe's *Faust* (Hayes 2010). At Poe's time, an allusion to Retzsch's devil would have thus most likely evoked the character Mephistopheles from Goethe's play. It may be fruitful to pursue this association further. If the man of the crowd is another Mephistopheles, then the narrator is, one could argue, another Faust – a man despairing over his failing pursuit of knowledge. Poe's "The Man of the Crowd" presents the Faustian tragedy of the modern metropolis. His story shows the limits of knowledge (and the failure of observation) under the specific conditions of the new urban environments.

Immediately after Poe's narrator has been so extraordinarily struck by the appearance of the man of the crowd, he senses "a craving desire to keep the man in view – to know more of him" (Poe 1965, vol. 4, 140). It is this desire to keep the curious man in view that marks the attempted transition from image to sequence – from the momentary glance to the ongoing surveillance that can render a narrative:

> Hurriedly putting on an overcoat, and seizing my hat and cane, I made my way into the street, and pushed through the crowd in the direction which I had seen him take; for he had already disappeared. (Poe 1965, vol. 4, 140)

The narrator's hurried departure from the protected lookout point behind the bow window of the coffeehouse has been repeatedly discussed in the scholarship (esp. by Sweeney 2003, 8–9; see also Kennedy 1975, 188). Readers have seen in this departure the decisive reason for the narrator's failure to gain any knowledge about the curious man who struck his attention. In an often-cited essay, Gerald Kennedy argues that the narrator could produce knowledge only as long as he remained behind the glass window, which distanced him from his objects of study (Kennedy 1975, 188). Susan Elizabeth Sweeney goes so far as to read Poe's "bow window" as a convex lens, prefiguring the mediating magnifying glass that Arthur Conan Doyle introduced some fifty years later in his first Sherlock Holmes novel, *A Study in Scarlet* from 1887 (Sweeney 2003, 7).[24]

What Kennedy and Sweeney do not acknowledge, however, is that the knowledge that Poe's narrator strives to attain at the moment when he decides to leave the coffeehouse is of a different kind than the knowledge of classification, which he pursued previously and which failed when confronted with the strange appearance of the man of the crowd. The failure of the classificatory system to account for the man of the crowd provokes the attempt to know this man's very own story. Struck by the curious features of the man of the crowd, the narrator is no longer satisfied with the knowledge of classification. Instead, he wants to know the "wild [...] *history* [that] is written within that bosom" (Poe 1965, vol. 4, 140, my emphasis). Instead of an abstract identification of the stranger's class status, he wishes to know the individualizing narrative, or "history," that constitutes the stranger's life. In order to find out about the stranger's history, the narrator has to "keep that man in view" and thus to leave his secluded lookout point.

---

24 It is from a distance and through the mediation of the glass, Sweeney and Kennedy argue, that the narrator manages the classification of the bypassing crowd. To substantiate their argument, Kennedy and Sweeney point to another of Poe's characters, the detective Dupin, who solves his cases from the secluded space of his own apartment (see Kennedy 1975; Sweeney 2003, 11; Poe's three detective stories featuring Dupin were published in the years after "The Man of the Crowd"). The bow window in "The Man of the Crowd" has been alternatively associated with a panorama (Sweeney 2003, 5) and, more than once, with a mirror (Auerbach 1989, 30, Kasson 1990, 85).

If the narrator ultimately fails to gain the individualizing observational knowledge that he sought, it is not because he leaves his secluded lookout point. It is, instead, because of the peculiar conditions of spectatorship that enable and limit the surveillance of strangers in the big city. The surveillance of strangers in the city becomes possible and necessary because one is surrounded by a mass of people that one can and must scrutinize in order to orient oneself among it. This orientation among strangers, however, can only work by means of classification, that is by quickly assigning each new person to a pre-existing set of categories or social types. While classification, thus defined, arguably describes an important initial aspect of all knowledge and experience, the big city limits the production of knowledge to this initial step because of the steady flow of new people who need to be categorized. The best urban spectator is simply he or she who knows the most categories and is fastest in assigning any new object of surveillance to one of these categories. As John F. Kasson has shown in his study *Rudeness and Civility: Manners in Nineteenth-Century Urban America*, much of nineteenth-century popular and specialized writing – from detective fiction to etiquette books and sociological atlases – works at formalizing precisely such a knowledge of classification in the urban environment.

"The Man of the Crowd" highlights the epistemological limits of this urban knowledge of classification. The expert of classification cannot see anyone isolated from his or her adherence to a larger class. The man of the crowd, who so struck the narrator's curiosity and whom the narrator sets out to follow, never leaves the crowd in whose midst he first appeared to the narrator. The narrator follows him for twenty-four hours; he paces with him up and down the thoroughfare in front of the coffeehouse, and through narrow streets to a crowded square. From there, they proceed to a marketplace and further. Every time one of the places empties out, the man of the crowd moves on to find new aggregates of the urban population. At one point, he seeks shelter among the audience that is rushing out of a closing theater. Later still, when most of London is asleep, he visits a bar in the poor outskirts of the city. Never does the narrator get a chance to see the man of the crowd alone.

The strange man's persistent adherence to the crowd can be read as an expression of the peculiar conditions of spectatorship in the city, which do not allow to view any individual categorically divorced from the mass as whose member he/she comes into view in the first place.[25] For the urban spectator,

---

25 J. Gerald Kennedy suggests, instead, that the strange behavior of the man of the crowd is due to the narrator's pursuit of him. Kennedy writes that "[a]n obvious explanation for the man's singular conduct (and his affinity for crowds) is his awareness of the narrator's presence" (Ken-

there is no possible form of knowledge outside the system of classification, which the man of the crowd's idiosyncratic expression escapes. The urban spectator is unable to isolate a person from the crowd in which he or she appears. This inability to focus fundamentally precludes the urban spectator from becoming an observer.

With the failure of observational knowledge in "The Man of the Crowd," the narrative itself fails too. The quasi-narrative of the quasi-observation in the English capital does not provide a real story to speak of – other than the story of observation's and narration's failure. We never pass from the description of an image to the narration of an extended sequence because the initial image always remains too blurry, lacking any identifiable individual form. Observation – both as a process of perception and as a literary technique that performs this perceptual process – collapses in "The Man of the Crowd" in the face of the all-dominating system of surveillance as classification. This problem of (literary) observation in an age of classification remains, I argue, an important theme in the literature of the following decades – not least in Arthur Conan Doyle's Sherlock Holmes stories, which I discuss in the final chapter. At the same time, however, Doyle's stories also present a new form of observation that appears to offer a solution to the difficult transition from the seeing of images to the seeing of sequences that plague Goethes's, Büchner's, and Poe's texts. My goal in the final chapter will be to analyze both, the promise of a new form of observation in Doyle's stories, and their remaining struggle with the problem of classification.

---

nely 1975, 190). However, we are not given any information in the text that would allow us to state with any certainty that the man of the crowd is indeed aware of being watched.

Chapter 5: Another Form of Observation?
(*Sherlock Holmes*)

In its countless episodes, Rétif's novel *Les Nuits de Paris, ou le Spectateur-nocturne* repeatedly moves from the description of a striking image to the sustained narration of a sequence of events in which this image is explained. This literary shift from description to narration performs and reflects a perceptual shift between two forms of seeing: the seeing of an unexpected image and the ensuing focused watching of a sequence of events. But Rétif pays little attention to the problems that lie dormant in the transition from the seeing of an image to the seeing of a sequence. Just how problematic this transition really becomes in a number of different cultural and literary contexts, we have seen in the readings of Goethe's *Werther*, Büchner's *Lenz*, and Poe's "The Man of the Crowd" in the preceding chapter. These texts, I argued, question the possibility to focus on a single image ("The Man of the Crowd"), to understand the ways in which an image changes over time (*Werther*), and to integrate the series of changing images in a coherent sequence (*Lenz*).

In this chapter, I argue that Arthur Conan Doyle's Sherlock Holmes stories, published between 1887 and 1927, provides us with a new form of observation that offers, in some sense, a solution to the problematic transition from the seeing of an image to the seeing of a sequence by suggesting that the image itself contains legible traces of a sequence of events. Sherlock Holmes, according to Watson's portrayal of him, observes the history of the crime *in* the crime scene. Sherlock Holmes practices a way of seeing that does not rely on the observation of events as they unfold over time; instead, Holmes's form of observation – if it is observation indeed – reconstructs series of past events from the presently visible and describable image of the crime scene. Image and sequence, description and narration collapse into one in Holmes's observations. As such, Holmes's acts of analysis offer a solution to the problems in the transition from image to sequence that Rétif failed to address. Moreover, Doyle's stories also stand in the most radical opposition possible to Lesage's story of devil and student over the roofs of Madrid, in which the description of the visible image and the narration of the story remained two entirely separate processes. It is, perhaps, not a coincidence that Doyle's detective series is also the last prominent literary work that directly references *Le Diable boiteux*. There really could not be any text that, while standing in the tradition of Lesage's text (of its focus on the visual), also questions its basic premises so entirely.

However, my reading of Sherlock Holmes's detective work as a crucial response to a two-century-long reflection on observational procedures in narrative

https://doi.org/10.1515/9783110594348-153

texts bears complication and qualification. Most importantly, we have to ask whether Sherlock Holmes's procedures truly qualify as observation. By this I mean not so much the narrative procedure of literary observation directly (i.e. the shift from description to narration), as the visual act (the act of seeing) that this literary procedure performs. This visual act of observation I understand with Lorraine Daston and Elizabeth Lunbeck to be centered on the openness "to possibilities for new knowledge in the most unexpected places" (Daston, Lunbeck 2011, 8). As I argued earlier in this book, the dual structure observation, which includes the seeing of images and the seeing of sequences, potential distractibility and sustained focus, is a product of observation's quest for new knowledge. Not knowing what they will find, observers must initially be open to being struck by any curious image. At the same time, however, this openness initially also limits the observer to the seeing of images, for the seeing of a sequence, which discloses a narrative development, would demand a prior focus on a given image. The observer must be able to perpetually go back and forth between the initial distractibility to be struck by any curious image and a more focused seeing that allows him/her to discover and record temporally extended sequences. My contention regarding Sherlock Holmes is that, contrary to what Watson suggests in his portrayal of the great detective, the narratives that Holmes constructs when studying an object are not based on proper observations in this sense. Holmes is rarely truly willing to be struck by a new image, and the crimes with which we are presented are generally repetitions of previous cases that Holmes recognizes to be identical to the present case. Thanks to Holmes's acute registration of the smallest clues and thanks to his vast index of older cases, Holmes can link any new case to a series of identical older cases. Sherlock Holmes, in other words, does not observe; he merely registers the clues and markers that allow him to draw the right stories from his database. The cases that Holmes solves can appear as curious new stories only to the faithful readers of Watson; for Watson fails to recognize these cases for what they fundamentally are: mere repetitions of an existing register of criminal cases. Through Watson's narrative, Holmes's mechanical registration of clues and markers is reinterpreted as observation. This tension between Watson's interpretation of the cases as curious, new, individual stories on the one hand, and the actual mechanics of Holmes's work on the other hand is crucial to the stories. While Holmes can solve his cases only if they are repetitions of older cases, his cases are worthy of storytelling only if they appear to be of singular interest. Doyle's stories confront us thus once more with the relation between classificiation, observation, and storytelling, which was, in different ways, already at stake in Poe's "The Man of the Crowd."

The fundamental ambiguity between the rhetoric of individuality (or new-ness) and the practice of classification, which is characteristic of *Sherlock Holmes*, has been ignored in much of the scholarship – most importantly per-haps in Carlo Ginzburg's influential essay "Morelli, Freud and Sherlock Holmes: Clues and Scientific Method" (1980), which shaped the scholarly discussion of Doyle's detective stories. At the core of Ginzburg's essay stands the problematic characterization of a range of distinct late-nineteenth-century cultural figures, including Sherlock Holmes, as all dealing with individualizing observation. Through a critical reexamination of the figures analyzed by Ginzburg, I show in this chapter that many cases in which Ginzburg sees individual observation represent in fact merely classificatory approaches, and I situate Doyle's Sherlock Holmes stories ambiguously in between observation and classification.

Sherlock Holmes observes the crime scene

In a famous scene near the beginning of the second Sherlock Holmes novel, *The Sign of Four* (1890), Sherlock Holmes analyzes Watson's watch. Holmes carefully studies the watch: he "balanced the watch in his hand, gazed hard at the dial, opened the back, and examined the works, first with his naked eyes and then with a powerful convex lens" (Doyle 1953, 92). Unlike the kind of observation in which Rétif's nocturnal spectator engages, the goal of Holmes's observation is not to capture the working of the watch over time. Looking at the watch, Holmes does not see time pass. The time that Holmes captures is, instead, the past time. Observing the watch, he studies its history and the history of its pre-vious owner, Watson's recently deceased brother. Holmes remarks correctly that the watch must have been passed on from Watson's father to Watson's brother before Watson himself received it, and he is also able to narrate much of Wat-son's brother's unhappy life:

> He was a man of untidy habits,—very untidy and careless. He was left with good prospects, but he threw away his chances, lived for some time in poverty with occasional short inter-vals of prosperity, and finally, taking to drink, he died. (Doyle 1953, 92)

Consider here the remarkable parallels and differences between Doyle's Sherlock Holmes, Rétif's nocturnal spectator, and, more importantly still, Lesage's devil Asmodée and student Don Cléofas. Looking at the present tableaux under the roofs of the houses of Madrid, Lesage's devil claims that it is necessary to know the respective backstory of the current moment; but he also states that only his superior knowledge of all things past can supply these stories. The presently visible tableaux themselves do not contain any legible traces of their stories – even though they are fully determined by the past events that are captured by these stories. This is completely different in the Sherlock Holmes novel. Holmes manages to tell the backstory to the presently visible tableau through the analysis of the tableau itself. And unlike Rétif's nocturnal spectator, the explanation of the present tableau does not work by recording over time the development that the tableau undergoes. Instead, Holmes claims that the story that explains the present tableau is contained in the present tableau and is only waiting to be deciphered. While Lesage completely divorces the description of the presently visible tableau from the narration of the prior story, and while Rétif defines the seeing of the describable image and the seeing of the narratable sequence as two separate steps in the observational process, Doyle has description and narration coincide in Holmes's observations.

In the Sherlock Holmes series, the divine omniscience of Lesage's devil is thus reconfigured as Holmes's rationalized observational procedure. Holmes manages to observe in the tableau (in the present example: in the watch) what the devil knows only qua being a devil. Doyle himself was likely aware of the fact that his detective stories constituted something like the modern retelling of Lesage's novel. At the very beginning of the early Holmes story "A Case of Identity,"[1] Doyle has his master detective prominently evoke Lesage's novel:

> "My dear fellow," said Sherlock Holmes as we sat on either side of the fire in his lodgings at Baker Street, "life is infinitely stranger than anything which the mind of man could invent. We would not dare to conceive the things which are really mere commonplaces of existence. If we could fly out of that window, hand in hand, hover over this great city, gently remove the roofs, and peep in at the queer things which are going on, the strange coincidences, the plannings, the cross-purposes, the wonderful chains of events, working through generations, and leading to the most *outré* results, it would make all fiction with its conventionalities and foreseen conclusions most stale and unprofitable." (Doyle 1953, 190 – 191)

The reference to *Le Diable boiteux* in this passage is unmistakable. In Holmes's imagination, he flies over London with Watson just as Lesage's devil Asmodée flies over Madrid with the student Don Cléfoas. To my knowledge, Doyle is the last prominent writer who evokes Lesage's novel and its mode of aerial spectatorship.[2] Moreover, as I showed in the analysis of the history of French editions of *Le Diable boiteux* (see Fig. 5) as well as of English translations of the novel, the popular reception of Lesage's novel comes to an end almost exactly at the time

---

1 My reading of "A Case of Identity" draws, in part, on an earlier article of mine (Wagner 2017).
2 See my remarks on the reception history of *Le Diable boiteux* in chapter 2.

when Doyle writes the Sherlock Holmes series. If *Le Diable boiteux* marks in some ways the beginning of an interest in literary procedures of observation, Sherlock Holmes marks, I argue, the culmination and end of this interest.[3]

One of the most important differences between Lesage's novel and the British mystery series lies, I submit, in the fact that Holmes imagines in his flight over London the possibility to see "wonderful chains of events, working through generations." As I discussed repeatedly in the preceding chapters, Lesage's protagonists do not observe such chains of events under the roofs of the houses of Madrid. In Lesage's novel, we almost only get to see momentary states of being; the stories themselves have to be provided by the devil. This is radically different in Doyle's detective stories. As I already noted in the example of Watson's watch, which contains the history of Watson's family, Holmes's observation of present objects allows us to see the history of events as they work literally "through generations."

There is a lot that could be said about the transformations in the culture of knowledge throughout the eighteenth and nineteenth centuries that allow for the fiction of Sherlock Holmes's observational skills to arise. Cultural historians will certainly want to highlight Walter Benjamin's analysis of the origins of detective fiction. Benjamin develops his argument about early detective fiction in the context of his discussion of the bourgeois ideal of building interiors in which one could inscribe one's identity as time passes – simply by living, and by leaving traces in these interiors. In his essay "Paris, die Hauptstadt des XIX. Jahrhunderts" [Paris, Capital of the Nineteenth Century], Benjamin explicitly links this bourgeois ideal to the phenomenon of literary detectives. The detective (re)constructs the perpetrator's identity from the visible traces in the crime scene and thus sustains the bourgeois belief that one can leave traces of a stable identity in the first place[4]:

> Das Interieur ist nicht nur das Universum, sondern auch das Etui des Privatmanns. Wohnen heißt Spuren hinterlassen. Im Interieur werden sie betont. Man ersinnt Überzüge und Schoner, Futterals und Etuis in Fülle, in denen die Spuren der alltäglichen Gebrauchsgegenstände sich abdrücken. Auch die Spuren des Wohnenden drücken sich im Interieur ab. Es entsteht die Detektivgeschichte, die diesen Spuren nachgeht. Die „Philosophie des Mobiliers" sowie seine Detektivnovellen erweisen Poe als den ersten Physiognomen des Interieurs. Die Verbrecher der ersten Detektivromane sind weder Gentlemen noch Apachen, sondern bürgerliche Privatleute. (Benjamin 1961, 178)

> The interior was not only the private citizen's universe, it was also his casing. Living means leaving traces. In the interior, these were stressed. Coverings and antimacassars, boxes and

---

3 I develop this argument in more detail in the conclusion.
4 On the construction of identity through the detective's work, see also Barloon 2006.

casings, were devised in abundance, in which the traces of everyday objects were moulded. The resident's own traces were also moulded in the interior. The detective story appeared, which investigated these traces. The *Philosophy of Furniture*, as much as his detective stories, shows Poe to have been the first physiognomist of the interior. The criminals of the first detective novels were neither gentlemen nor apaches, but middle-class private citizens. (Benjamin 1969, 166)

However, aside from this cultural analysis, one should also consider the ways in which contemporary scientific observation informs Holmes's detective work. For Holmes's expectation to find in the present scenes legible traces of a past sequence of events defines many nineteenth-century scientific observational endeavors. I will here just briefly gesture to two major figures of nineteenth-century science, Charles Lyell and Charles Darwin. The work of both men has repeatedly been linked to the narrative logic underlying the Sherlock Holmes stories (see, for instance, Frank 2003, 154–175, Kerr 2013, 126–127). Moreover, Doyle, who discussed in his Professor Challenger novels (notably in *The Lost World* [1912]) matters of geology and evolutionary biology, most likely knew Lyell's and Darwin's books. In our context, Lyell's work is especially important for its geological reconstruction of the long history of the Earth from currently visible traces.[5] Lyell's influential *Principles of Geology* (1830–1833) with its telling subtitle *Being an Attempt to Explain the Former Changes of the Earth's Surface, by Reference to Causes Now in Operation* very much emphasizes geology's achievement to have discovered the long history of our planet's dynamic development and thus to have broken with the wrong belief of men who had "for ages declared the earth to be at rest" (Lyell 1997, 24). Importantly, Lyell's reconstruction of the world's history relies on the patient description of the present state of the Earth – or, more precisely, of the minute processes of sedimentation and erosion now in operation, which have led over millions of years to the emergence and disappearance of entire continents.

Charles Darwin's first tentative formulation of the idea of evolution is similarly developed from the 'image' of the presently visible fauna. In the chapter of *The Voyage of the Beagle* (1839) that covers his sojourn on the Galapagos archipelago, Darwin famously describes the great variety among the species of finches inhabiting the different islands of the archipelago (Fig. 11). It is the minute description of the gradation now present in the sizes of the finches' beaks that leads Darwin to the assertion of a temporal sequence that gave rise to these dif-

---

5 On the transformations of the geological discourse of the nineteenth century, see, more generally, Rudwick 2010.

ferent forms. He thus constructs a narrative of development from the image of contemporary variety:

> Seeing this gradation and diversity of structure in one small, intimately related group of birds, one might really fancy that from an original paucity of birds in this archipelago, one species had been taken and modified for different ends. (Darwin 1997, 361)

Doyle's detective work participates in this new scientific episteme according to which long histories of preceding events remain legible in the present state and only wait for the observer to be deciphered.[6]

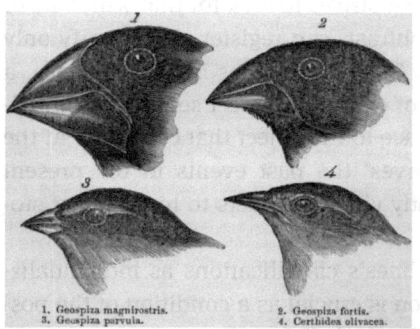

Figure 11: Different species of finches on the Galapagos archipelago; illustration from Charles Darwin's *The Voyage of the Beagle.* Reproduced with permission from John van Wyhe, ed. 2002-. *The Complete Work of Charles Darwin Online.* (**http://darwin-online.uk/**).

Observation or classification?

But while Holmes's analyses, on one level, indeed seem to correspond to contemporary scientific procedures of observation, they ultimately fall short of the open-ended production of knowledge that is at the basis of scientific observation. Observation, as Lorraine Daston and Elizabeth Lunbeck insist, is defined by the observer's openness "to possibilities for new knowledge in the most unexpected places" (Daston, Lunbeck 2011, 8). The observer does not know which knowledge he/she will gain through his/her observation. The fundamental openness of the observer to new knowledge necessitates the dual structure of observation as the combined openness to seeing unexpected images and the sustained focus on extended sequences. Because observers do not know at the outset of their study what they will find or what they should focus on, they re-

---

6 In his new monograph on Doyle, Douglas Kerr explicitly links the scientific work of Darwin and Lyell to the scene of Holmes's observation of Watson's watch at the beginning of *The Sign of Four* (Kerr 2013, 127).

main initially limited to the seeing of momentary images. Subsequently, however, the observer focuses on the one image that struck his/her attention (while remaining, to some degree, ready to be distracted anew).

My thesis is that the procedures that figure as "observation" in Sherlock Holmes consist, in fact, in mere acts of mechanical registration and classification. In contrast to observation, classification does not evolve over time and does not produce new knowledge based on individual cases. Instead, classification consists in the mere reapplication of an already existing system of knowledge on new bits of data. A traditional computer, were it fed with sufficient information, could classify everything according to pre-established categories, but it could not be made to observe. The observer strives to look for that which he or she does not know yet; a computer, by contrast, can register and classify only that which it categorically already knows. Sherlock Holmes is, in this sense, a computer. He possesses a vast knowledge of cause-and-effect sequences that allows him to link almost any given visual trace to the object that caused it. At the points at which Holmes seemingly 'observes' the past events in the present scenes, he, in fact, only relates the presently visible markers to his index of stories (of cause and effect).

Watson perpetually misrepresents Holmes's classifications as individualizing observations, and this misrepresentation is crucial as a condition of the possibility of storytelling. As literary stories, Sherlock Holmes's cases are only ever possible as singular adventures and individual observations. Why else should we be interested in them? At the same time, however, Doyle rarely misses a chance to hint at the fact that we really read and experience only the reproduction of an old play. This subliminal triteness of the cases at hand is essential to them as a condition of Holmes's success. Holmes is able to solve the cases only because he has already experienced identical cases before and now sees the same phenomena reappear.[7]

---

7 This is the case at least for a good number of Doyle's Sherlock Holmes stories. I disagree therefore with those readers who believe that Sherlock Holmes works most importantly by way of a Searlian "abductive reasoning," i.e. by informed guesses that always need further confirmation. Contrary to what Thomas Sebeok and Jean Umiker-Sebeok claim, many of Holmes's achievements are not due to his "extraordinary capacity for sustaining the guessing-testing-guessing chain" (Sebeok and Umiker-Sebeok 1979, 23). While Eco and others may be right in claiming that real-world detectives – and also many fictional detectives – rely precisely on abductive reasoning and chains of hypothesis-building and testing (see Eco 1983), Sherlock Holmes's superior knowledge makes such complicated procedures unnecessary. As J. G. Kennedy has already critically noted in his review of Eco and Sebeok's influential collection of essays *The Sign of Three: Dupin, Holmes, Peirce*, Holmes, in contrast to Peirce's real-world-reasoner, does not op-

We are thus confronted again with the problem of classification with which we were already concerned in the reading of Poe's "The Man of the Crowd." Poe's narrator, the prototypical modern urban spectator, remains limited to classification. His competence in classification enables him in the beginning of the story to sort the crowd in front of the coffeehouse in London. He distinguishes, for instance, the upper clerks from the lower clerks, and the different divisions among the urban poor. Every person walking by the coffeehouse belongs to one of the categories in the narrator's system of classification. However, classification does not produce individual stories. On the contrary, it presupposes that all stories have already been told. In the classificatory system, we can only say: he or she is *also* one of those who…

Once Poe's narrator stumbles upon a figure that defies easy classification – i.e. the mysterious "man of the crowd" – his faculties are paralyzed. While the man of the crowd cannot be classified in any unambiguous way, he cannot be watched as the singular protagonist of his very own story either. The narrator never manages to divorce the mysterious man from the crowd in which he first appeared. Much of the 'story' of "The Man of the Crowd" evolves in the nowhere land between classification and individual observation. Poe's narrator remains unable to tell a (conventional) story because he is unable to individualize the man of the crowd in a sufficient way to make him the subject of his very own story.

In Doyle's Sherlock Holmes series, the same problem of classification, observation, and storytelling recurs – with the crucial difference, however, that Doyle keeps the problem at bay by introducing a narrator who fails to recognize classification for what it is. Watson presents as individualizing observation what is in fact mere classification. He tells individual stories where there is actually only repetition.

Consider once more the example of Holmes scrutinizing Watson's watch. Holmes is able to see the history of the brother's alcoholism inscribed on the watch only because he has seen the connection between the markers on the watch and a specific biographical narrative in precisely this way many times before. Holmes explains:

> Finally, I ask you to look at the inner plate, which contains the key-hole. Look at the thousands of scratches all round the hole,—marks where the key has slipped. What sober man's key could have scored those grooves? *But you will never see a drunkard's watch without*

---

erate under the conditions of fallibility (Kennedy 1986, 123; Eco partially concedes this [Eco 1983, 219–220]).

them. He winds it at night, and he leaves these traces of his unsteady hand. *Where is the mystery in all this?* (Doyle 1953, 93, my emphasis)

As Holmes makes clear, a drunkard's watch always looks like the one at hand. Holmes's final question – "where is the mystery in all this?" – is symptomatic of the kind of mystery novels that Doyle writes. From a certain perspective, there really is no mystery in them. These stories are singular mysteries only from the perspective of him or her who fails to recognize them for what they are: repetitions and reapplications of an already existing system of classification. The application of this system does not require any sustained observations, but only a decipherment of the respective markers. To be sure, Holmes refers to his own procedures at numerous points as "observations." However, "observation" for Sherlock Holmes means merely to register any and all possible markers. Holmes's "observation" does not produce new knowledge in places where he did not expect to find it, and it is, by implication, not characterized by the combined seeing of image and extended sequence, which I analyzed in *Les Nuits de Paris*. True observers proceed from completely unfocused spectatorship in which they are open to being struck by any image that strikes their curiosity, to the sustained watching of a sequence that discloses the history or meaning of the observed image. Holmes, by contrast, is neither ever wholly unfocused, nor completely focused. Instead, he systematically scans objects for markers that fit his categories.

There certainly exist procedures of reconstructing crimes from the study of the crime scene that actually qualify as observations. Michelangelo Antonioni's 1966 film *Blow Up* about a young photographer who witnesses a crime is an interesting example. Struck by some irregularity in a picture that he recently took, Antonioni's photographer keeps enlarging the photograph until he sees that the image seems to bear witness to a murder that was possibly being committed at the moment when the photo was taken. Through the enlarged versions, the initial photograph is shown to contain traces of the sequence in which the murder was executed. Later on in the film, the reliability of the evidence of these enlarged photos is increasingly called into question. But the question of evidence, which is important to Antonioni's film, is secondary to the present study of observation. Antonioni's photographer is, according to my definition, a true observer because he proceeds from the registration of an unexpected momentary image to the sustained watching of a sequence – even though this sequence is contained in the image itself. Sherlock Holmes, by contrast, approximates such observation as it is contained in *Blow Up* only seemingly. In contrast to Antonioni's photographer, Sherlock Holmes can easily connect the strange details of a present image (of the crime scene) to an archive of explanatory narratives.

Much of the corpus of the Sherlock Holmes stories is characterized by the tension in which Holmes's actions are portrayed simultaneously as singular observations and as mechanical registration and application of a pre-existing body of knowledge. This tension is personified in the tension between Holmes – a true dandy in Baudelaire's sense: a man who has seen it all – and Watson, who, although he is a veteran and practicing doctor, comes to the world like a newborn baby, shocked by every little feat of his friend Holmes, trite and predictable as it may be.

The tension in the representation of Holmes's cases as observed singularity and registered repetition is particularly prominent in the story "A Case of Identity," at which I take a closer look in the following pages. Strikingly, this is also the first and only Sherlock Holmes story that includes the word "case" in its title. In contrast to the majority of the stories, which are called "adventures" and thus suggest singularity, this 'case' prepares us from the outset for the problem of a seriality (of cases). The idea of seriality is further underscored by the precise title, "*A* Case of Identity," suggesting that there is more than just one such a case. This, we learn later on in the story, is in fact true. And because there are other parallel cases, Holmes knows the solution to the present case immediately after his client, Miss Sutherland, has presented her problem. Holmes remarks to Watson:

> "I found her [Miss Sutherland] more interesting than her little problem, which, by the way, is rather a trite one. You will find parallel cases, if you consult my index, in Andover in '77, and there was something of the sort at The Hague last year." (Doyle 1953, 196)

As Holmes makes clear, there are really already some (almost) identical cases to this "Case of Identity" in Holmes's "index[ed]" collection, and this makes it easy to solve the present case.

To be sure, these older cases are only "parallel" and "something of the sort." They are not entirely identical to the present case. But this should not surprise us. For the epistemology of cases suggests not only seriality, but also a lingering irreducible singularity. Cases are never completely identical to each other. For Holmes, however – and this is crucial – these remaining differences are not important. In fact, his analytic strength consists precisely in his ability to abstract from these differences and to recognize the structural "identity" of the cases.

It does not take Holmes any careful observation to solve the case at hand; he merely has to register that the case is identical to other cases in his existing collection. Holmes only has to "consult [his] index." This index is mentioned in several of the Sherlock Holmes stories. In "A Scandal in Bohemia" (1891), the very first Sherlock Holmes short story, published briefly after the initial two novels (*A Study in Scarlet* [1887] and *The Sign of Four* [1890]), Watson explains:

> For many years he [Sherlock Holmes] had adopted a system of docketing all paragraphs concerning men and things, so that it was difficult to name a subject or a person on which he could not at once furnish information. (Doyle 1953, 165)

This all-encompassing index, whose capacity puts modern-day electronic search engines to shame, constitutes the condition of possibility of Holmes's detective skills.

Holmes's expressed familiarity with the new case in "A Case of Identity," I now wish to show, should be understood as an important element of a sustained poetological and epistemological debate in this early Sherlock Holmes story. Interestingly, the roles in this debate are initially different from what one would expect. Sherlock Holmes, whose actual detective work relies on a recurrence of similar cases, appears here at first as the proponent of a structure of singularity in life. Watson, who in his role as narrator generally misinterprets Holmes's stories as singular adventures, here initially questions the existence of truly singular events in life.

The debate between Holmes and Watson (as well as the entire story) starts with Holmes's claim that "life is infinitely stranger than anything which the mind of man could invent" (Doyle 1953, 190). As I indicated, this claim about the strangeness of reality seems to be in stark contradiction to Holmes's reliance on the reoccurrence of identical cases in his actual detective work.[8] To make things even more confusing, Watson contests Holmes's claim and argues instead for the repetitive structure of reality. The two characters thus each assume the opposite position from the one that they normally occupy. This riddle is solved, however, if one recognizes that, over the course of the story, they switch their positions again. At the end, Holmes reveals his reliance on the repetition of identical cases while Watson is struck by Holmes's singular finding. The story thus

---

8 There is possibly a way to solve the apparent contradiction between Holmes's initial claim for the superiority of reality over fiction on the one hand and his subsequent testimony to a completely indexed reality on the other hand. Reality may be repetitive in structure, but its repertoire can be still more diverse and interesting than the restricted patterns of fiction. As Holmes makes clear, there are commonplaces in reality as well: "we would not dare to conceive the things which are really mere commonplaces of existence" (Doyle 1953, 190–191). However, this attempt to solve the tension between the depiction of reality as simultaneously repetitive and strange remains problematic. Holmes, for instance contradicts his own statement about the "commonplaces of existence" immediately afterwards by claiming that "nothing is so unnatural as the commonplace" (Doyle 1953, 191).

stages the process in which Holmes and Watson develop their fundamental positions regarding singularity in reality. "A Case of Identity," it could be said, presents something like the epistemological myth of foundation of the Sherlock Holmes stories. In the following, I will reconstruct how this myth proceeds to present us in the end with the familiar positions of Holmes and Watson.

At the beginning of the story, Holmes sets out to prove his claim about the structure of singularity of reality through the thought experiment of a flight over London, imagining what he could see if only he had the chance to lift the roofs of the houses and peep in (I quoted the respective passage earlier in this chapter). Needless to say, Holmes pursues a strange strategy of argumentation at this point. For the fiction of the flight over London seems to speak just as much to the power of imagination as to the power of real vision.[9] Watson does not comment on this tension – and possible contradiction – in Holmes's argument. He does, however, contest the claim that observation would truly reveal singularity, and he cites as evidence the trite cases that are published daily in the newspapers:

> "And yet I am not convinced of it," I answered. "The cases which come to light in the papers are, as a rule, bald enough, and vulgar enough. We have in our police reports realism pushed to its extreme limits, and yet the result is, it must be confessed, neither fascinating nor artistic." (Doyle 1953, 191)

Holmes, however, insists that "nothing [is] so unnatural as the commonplace" (Doyle 1953, 191), and it makes good sense to read the rest of the story also as a test case for Holmes's claim. Holmes and Watson agree on a wager on the singularity of real events, as it were, and the ensuing story must decide who of them wins the wager.

At first instance, Holmes seems have the upper hand. In order to prove the triteness of reality, Watson "picked up the morning paper from the ground" and suggests: "let us put it [Holmes's claim about singular reality] to a practical test" (Doyle 1953, 191). Watson subsequently reads out the headline of the first article he stumbles upon: "Here is the first heading upon which I come. 'A husband's cruelty to his wife.'" (Doyle 1953, 191) Without reading any further, Watson already believes to know what is to follow and predicts the chain of events himself:

> "There is half a column of print, but I know without reading it that it is all perfectly familiar to me. There is, of course, the other woman, the drink, the push, the blow, the bruise, the

---

9 I have discussed the discourse on fictionality in the Sherlock Holmes stories (with special attention to "A Case of Identity") more extensively elsewhere (Wagner 2017).

sympathetic sister or landlady. The crudest of writers could invent nothing more crude." (Doyle 1953, 191)

Watson's strategy is, just as Holmes's before, deeply flawed. Instead of reading the case in the newspaper, he imagines the case himself. His prediction could therefore only ever support Holmes's point about the triteness of imagination; it could not prove the triteness of the case itself, which remains to be studied. Just as Holmes's argument in favor of observation was based on the imagination of the flight over London, Watson's argument against real cases is based on imagination as well.

As it turns out, the case upon which Watson stumbles in the newspaper is of a more singular nature than Watson had expected. Holmes, who claims to know the case personally, can tell Watson about its most striking features:

> "Indeed, your example is an unfortunate one for your argument," said Holmes, taking the paper and glancing his eye down it. "This is the Dundas separation case, and, as it happens, I was engaged in clearing up some small points in connection with it. The husband was a teetotaler, there was no other woman, and the conduct complained of was that he had drifted into the habit of winding up every meal by taking out his false teeth and hurling them at his wife, which, you will allow, is not an action likely to occur to the imagination of the average story-teller." (Doyle 1953, 191)

Indeed, what could Watson possibly say after Holmes has "hurled" this strange case at him – not unlike the idiosyncratic husband in the story, who hurled his teeth at his wife?

What Watson does not do, in any case, is read the newspaper article and confirm whether Holmes's version is accurate or merely the product of the detective's imagination. Holmes does not let it come to that. Having finished his story, Holmes underscores his victory in the wager and puts an end to the conversation: "Take a pinch of snuff, Doctor, and acknowledge that I have scored over you in your example." (Doyle 1953, 191) The snuffbox that is presented here further strengthens Holmes's point. It is a "snuffbox of gold, with a great amethyst in the centre of the lid" (Doyle 1953, 191). The box is a recent present from the king of Bohemia, Holmes's client in the story "A Scandal in Bohemia," which was published briefly before "A Case of Identity."[10] The extraordinary snuffbox underscores the extraordinariness of Holmes's real-world cases, and, along with it, potentially also the overall structure of singularity in reality.

---

10 "A Scandal in Bohemia" is also the first story in the volume *The Adventures of Sherlock Holmes*, in which "A Case of Identity" appeared as well.

However, the initial newspaper story is actually only the prelude to the real test case that is laid out in "A Case of Identity." And at this second instance, Holmes loses the wager. More precisely, the point is that he *has to* lose in order to solve his case successfully. As a rule, he can solve his cases only if they fit his index of classified chains of cause and effect.

Having discussed the Dundas separation case, Watson asks his friend whether he is presently working on any case. Holmes answers in the affirmative – he is working on "[s]ome ten or twelve" (Doyle 1953, 191) – but readily admits that that "none [of these] present any feature of interest" (Doyle 1953, 191). While the long list of Holmes's boring cases does not bode well for Holmes's argument about singularity in reality, Holmes already hopes to be vindicated by the case into which he is about to enter. Spotting a woman outside on the street, he believes to see a future client. Holmes conjures: "It is possible, however, that I may have something better before very many minutes are over, for this is one of my clients, or I am very much mistaken." (Doyle 1953, 192) Gazing out of the window of his apartment, Holmes evokes – albeit in an inversion – the initial fantasy of the observational flight over London and the examination of the interiors of the houses, and thus, by implication, also the wager with Watson.

The question that we have to ask ourselves is, once more, how strange are the things that Holmes and Watson see when they look out of the window. As Watson joins Holmes at the window, he notices a

> large woman with a heavy fur boa round her neck, and a large curling red feather in a broad-brimmed hat which was tilted in a coquettish Duchess of Devonshire fashion over her ear. From under this great panoply she peeped up in a nervous, hesitating fashion at our windows, while her body oscillated backward and forward, and her fingers fidgeted with her glove buttons. Suddenly, with a plunge, as of the swimmer who leaves the bank, she hurried across the road, and we heard the sharp clang of the bell. (Doyle 1953, 192)

Holmes is right; the woman on the street is indeed a new client. As she walks up the stairs to Holmes's apartment, the detective explains his chain of reasoning:

> "I have seen those symptoms before," said Holmes, throwing his cigarette into the fire. "Oscillation upon the pavement always means an *affair de coeur*. She would like advice, but is not sure that the matter is not too delicate for communication. And yet even here we may discriminate. When a woman has been seriously wronged by a man she no longer oscillates, and the usual symptom is a broken bell wire. Here we may take it that there is a love matter, but that the maiden is not so much angry as perplexed, or grieved. But here she comes in person to resolve our doubts." (Doyle 1953, 192; emphasis in the original)

The problem with Holmes's being right, however, is that his predictions were founded on the assumption that his potential new client would behave in exactly the same way as all previous female clients behaved. His success through correct classification tends thus to undermine his prior claim about the singularity of reality. "Depend upon it, there is nothing so unnatural as the commonplace" (Doyle 1953, 191), Holmes had claimed at outset of the story. This woman's behavior, however, is completely foreseeable. But then again, it is precisely this predictability that allows for Holmes's analysis. Indeed, Holmes's skills "depend upon" the fact that nature is full of commonplaces and repetitions.

Holmes proves his skills of classification in the actual case as well. However, proving successful means, once more, disproving the claim about the singularity of life. The case at hand is particularly apt to highlight Holmes's detective system of classification. As the story is among the most well known in the canon, I will limit its recapitulation here to the bare minimum. Holmes's client, Miss Mary Sutherland, explains to Watson and Holmes that her stepfather, who was trying to keep her inheritance from an American uncle for himself, prevented her for a long time from meeting any man she might be tempted to marry. For if she married, the stepfather would no longer have access to her inherited money. Finally, however, Miss Sutherland did manage to meet someone, and she soon got engaged. But on the morning of the wedding, her groom suddenly vanished, leaving her behind with her promise always to remain loyal to him. Now she has come to Sherlock Holmes, hoping that the great detective will retrieve her lost fiancé.

While this case leaves Miss Sutherland and Watson completely puzzled, Holmes disentangles it immediately. (However, he discloses the solution to Watson only some time later, at the end of the story, so that there remains a mystery for Watson and the reader.) As we learn from Holmes in the end, Miss Sutherland's supposed fiancée is identical to her own stepfather, who had disguised himself as another man. He entered into the relationship with his stepdaughter with the sole intent to keep her from marrying anyone in earnest.

If Holmes manages to see the identity between groom and stepfather right away, he does so by quickly aligning the present case with a number of identical cases that are listed in his index. Recognizing the identity of stepfather and groom in the case thus works through the recognition of the identity of the present case with other, older cases. Holmes, then, is quite right to call "trite" (Doyle 1953, 196) a case that to Watson, who has no knowledge of similar or identical cases, remains a "singular mystery" (Doyle 1953, 198).

Although the initial conversation about the strangeness of life and the commonplaces of fiction is not taken up again at the end of "A Case of Identity," Watson must be convinced that Holmes was right: reality is stranger than anything he could have ever invented. Indeed, Watson has to depend upon the fact that reality is strange if his chronicling of Holmes's cases wants to make any claim on the reader's interest.[11] Holmes, by contrast, who played the advocate of reality's idiosyncrasy at the outset of the story, now seems to contradict his own statement. At least the present case lacks almost all singularity. In this way, "A Case of Identity" stages the process in which Holmes and Watson assume their complementary positions regarding singularity in reality.

Holmes's practice of 'observation' plays a crucial role in the construction of the cases as both singular and common. Observation in the Sherlock Holmes series is the practice in which the two contrasting perspectives on the case as common and singular are confused. When Holmes studies an object, the reapplication of the existing index is disguised in Watson's narrative as the individualizing production of knowledge through observation. The Sherlock Holmes series wavers between observation and classification. Depending on the perspective, Holmes appears either as the perfect observer or not as an observer at all.

This emphasis not only on individualizing observation but also on classification in Doyle's detective stories may appear surprising and historically almost unlikely. After all, the Sherlock Holmes stories were published toward the end of the century that had, in Michel Foucault's influential interpretation, invented the knowledge of the individual through a wide range of systems of discipline and control. Schools, prisons, hospitals, and psychiatry wards had worked at the development of a knowledge and language of the individual. As Foucault states with respect to the treatment of prisoners in the nineteenth century:

> Mais le plus important sans doute, c'est que ce contrôle et cette transformation du comportement [in prison] s'accompagnent—à la fois condition et conséquence—de la formation d'un savoir des individus. (Foucault 1975, 148)

> [N]o doubt the most important thing was that this control and transformation of behavior [in prison] were accompanied – both as a condition and as a consequence – by the development of the knowledge of the individual. (Foucault 1977, 125)

Given this emphasis on the idea that the nineteenth century developed a historically unprecedented knowledge of the individual, the assertion that individualizing observation in literature was overcome in the end by classification appears

---

11 It is interesting to note in this context that Holmes chooses not to disclose the identity of stepfather and fiancé to his client, Miss Sutherland. Just as Watson remains convinced of the singularity of the case, Miss Sutherland remains convinced of the actual individual (singular) existence of her fiancé.

as a strange anachronism. But the point of the matter is that at least some of the techniques to describe individuals actually rely very much on classificatory mechanisms. While this clarification may not come as a surprise to Foucault or his readers, it appears that in studies of Sherlock Holmes, this prevailing importance of classification has sometimes been overlooked.

This oversight is especially striking in Carlo Ginzburg's influential essay "Morelli, Freud and Sherlock Holmes: Clues and Scientific Method" from 1979 (English translation 1980).[12] In this essay, Ginzburg contrasts the suppression of the individual for the sake of measurable, repeatable procedures during the scientific revolution of the seventeenth century on the one hand, with the focus on the individual as an object of knowledge in the nineteenth century on the other hand. With respect to the earlier form of science, Ginzburg explains:

> Using mathematics and the experimental method involved the need to measure and repeat phenomena, whereas an individualising approach made the latter impossible and allowed the former only in part. (Ginzburg 1980, 15)

For thinkers like Galileo Galilei (1564–1642), who serves Ginzburg as an important example of the early modern scientist, the individual was beyond reach. For these thinkers, the individual has to be ignored in order to make science scientific Only if we ignore individual features can we begin to have systems of measurement and scientific laws. The individual escapes all measurement and all formulas.

According to Ginzburg's grand narrative, the nineteenth century developed new ways of including the individual as an object of knowledge. Ginzburg illustrates this quest for the individual at the outset of his essay in his analysis of Giovanni Morelli's art criticism, Sherlock Holmes's detective work, and Freud's psychoanalytic practice. Ginzburg highlights the striking commonality between these three seemingly so different figures: all three derive their results from the analysis of minute detail.

The problem with Ginzburg's analysis, however, is that he fails to pay sufficient attention to the different expectations with which Morelli, Freud, and Holmes approach detail. More precisely, Ginzburg overly relies on the concept of detail in his argument about singularity. Contrary to what Ginzburg suggests, detail and singularity are by no means synonymous. It would be perfectly possible, for instance, to study some detail in a given painting without having to claim that the respective detail is in any way unique to the painting. 'A typical

---

12 I refer in the following solely to the English translation, in which form this essay has had a great impact on the scholarship.

detail' is not an oxymoron. We can get a better understanding of this if we consider Ginzburg's example of the art critic Giovanni Morelli (1816 – 1891), who developed a method of assigning paintings to Italian renaissance masters by focusing on the way in which minor features such as fingernails or earlobes are painted. Morelli convincingly reasoned that imitators of the great masters were unlikely to copy faithfully these secondary aspects. Morelli thus looks at detail, but with a classificatory rationale: he creates patterns of earlobes or fingernails in the work of a given painter to compare them with the earlobes and fingernails on a given new painting. This allows Morelli to know, for example, if a painting is truly by Raphael. If the earlobes do not look like the earlobes on Raphael's other paintings, Morelli knows that the painting cannot be by Raphael.

Morelli's stratagem is markedly different from that of Freud. To be sure, Freud is also interested in previously ignored details: misspoken words, lapses of memory etc. But the way Freud works with these details is very different from Morelli. What is interesting in this context, however, is that not only Ginzburg, but also Freud himself overestimates the similarity between his (i.e. Freud's) work and that of Morelli. In his 1914 essay "Der Moses des Michelangelo" [The Moses of Michelangelo], Freud explicitly compares his method to that of Morelli:

> Ich glaube, sein [Morelli's] Verfahren ist mit der Technik der ärztlichen Psychoanalyse nahe verwandt. Auch diese ist gewöhnt, aus geringgeschätzten oder nicht beachteten Zügen, aus dem Abhub – dem *refuse* – der Beobachtung, Geheimes und Verborgenes zu erraten. (Freud 2000, vol. 10, 207, emphasis in the original)

> It seems to me that his [Morelli's] method of inquiry is closely related to the technique of psycho-analysis. It, too, is accustomed to divine secret and concealed things from despised or unnoticed features, from the rubbish-heap, as it were, of our observations. (Freud 1955, vol. 13, 222)

Freud's comparison of his method to that of Morelli certainly speaks strongly in favor of Ginzburg's claim that these different thinkers essentially follow the same epistemic assumptions. However, despite what Freud himself says about his relation to Morelli, his practice differs decisively from that of the art critic. This becomes especially clear in Freud's analysis of Michelangelo's Moses statue (c. 1515), which he presents in the very same essay that includes the reference to Morelli. Even though both Morelli and Freud look at details, Morelli does so with the goals of classification and identification: while he does look at details in the paintings, the systematic study of these details allows him in the end once more to identify recurring patterns with the help of which he can subsequently assign a new painting to the correct master. Morelli's method thus also resembles the technology of identification through fingerprints, which British authorities adopted in the last quarter of the nineteenth century and which figures in Ginzburg's essay as another example of the nineteenth-century study of singularity (Ginzburg 1980, 27). Once more, the valuation of fingerprints implies an emphasis on detail and speaks, in some limited sense, to the nineteenth-century endeavor to develop a knowledge of the individual. But the mechanism of identification through fingerprints is essentially classificatory: one establishes what the fingerprints of a given person look like and is subsequently able to identify this person again through the mere comparison to known fingerprints. This practice of classification with the help of hitherto ignored details is very different from the work that Freud undertakes in his analysis of the Moses statue.

It is true, Freud, too, explicitly focuses on previously overlooked details of the sculpture – "Details, die bisher nicht beachtet, ja eigentlich noch nicht richtig beschrieben worden sind" (Freud 2000, vol. 10, 207) [details which have hitherto not only escaped notice, but, in fact, have not even been properly described (Freud 1955, vol. 13, 222)]. The careful description of these details – namely the gesture of Moses's right hand and the position of the tables of the law (see Fig. 12) – leads Freud to reconstruct the movement that Moses must have completed before he reached the position in which the artist shows him. This, in turn, allows Freud to revise conventional interpretations of the sculpture.

According to the dominant opinion among art critics of Freud's time, Michelangelo shows Moses at the moment when he sees the adoration of the golden calf by his people and is about to rush up in anger and shatter the tables of the law. Freud, by contrast, insists that the position of beard, hands, and tablets forces one to assume a slightly different narrative and situation. Freud's reasoning and description are too intricate and nuanced to recapture here completely, but the basic summary is this: turning his head left, Moses had noticed the adoration of the golden calf; angrily he grabbed his beard with his right hand; this movement, however, caused the tablets under his arm to slip forward. Regaining his calm and trying to prevent the holy tables from being damaged, Moses pulls his right hand back; his beard, only partly released, follows this backward movement of the right hand: this is the situation captured by Michelangelo. To illustrate the movement that Freud ingeniously reconstructs in his careful observation of the statue, Freud includes in his essay a series of three drawings that he had commissioned from an artist. These drawings, which together look almost like a comic strip, show the successive stages of Moses's movement, first being entirely at rest, then being enraged by the adoration of the golden calf, and finally moving back to his previous position (Fig. 13).

Freud's reconstruction of the movement leading up to the presently visible scene is strikingly different from the analysis of detail that Morelli undertakes. Freud does not show the correspondence of any detail of the Moses statue to

Figure 12: Michelangelo, *Moses*, c. 1515, Rome, San Pietro in Vincoli. Photo by D. Anderson. Copied from Sauerlandt 1911, 13.

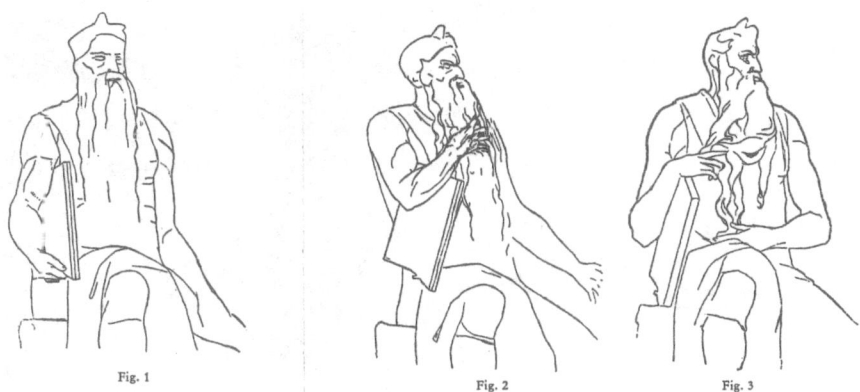

Fig. 1   Fig. 2   Fig. 3

Figure 13: Drawings commissioned by Freud to illustrate Moses's movement. Copied from Freud 2000, vol. 10, 212–213.

an existing index of similar details. Instead, he truly observes the individual details to read in the present scene the traces of its prehistory. One could, to be sure, critically remark that what Freud does in this essay is almost an exception to his general psychological work in which – or so some of his critics say – he tends to quickly apply the ever-same Oedipal matrix to every new individual case.[13] But maybe one could more neutrally suggest that the kind of observation of detail that Freud showcases in his analysis of Michelangelo's *Moses* serves as an ideal – for Freud as well as for his contemporaries. This is at least what the Sherlock Holmes stories would indicate. As I argued above, Watson's narrative perpetually works at the reinterpretation of a Morelli-like classification of detail as Freudian individualizing observation. In Watson's narrative, Sherlock Holmes appears as the perfect observer who can reconstruct from the careful analysis of the crime scene the history of the past crime. Watson shows us Sherlock Holmes as someone who can see in the present image the sequence of events that led to the present image. But through this thin veneer of perfect observation, a new reality of classification shines through.

---

13 On the critique of Freud's method, see, e.g., Crews 1995.

Conclusion: Literary Observation after 1900

Doyle's detective stories mark, in my understanding, an important historical and logical endpoint to the development and scrutiny of literary observation. However, as with most proclaimed endings in history, the 'end' of literary observation is neither absolute nor unambiguously locatable. Therefore, I will devote a few of the remaining pages in this book to the productive afterlife of literary observation in the twentieth and twenty-first centuries. But before I arrive at this afterlife of observation, I first want to establish that there was a rupture at all.

Historically, the Sherlock Holmes stories stand at an endpoint of literary observation in at least three distinct ways. First, Doyle's detective stories coincide with the end of the age of observation in the sciences: since the mid-nineteenth century, planned experimentation was increasingly seen as superior to the open-ended endeavor of observation in nature. Moreover, as Martin Jay has shown in his history of "ocularcentrism" in the Western World, the reliance on vision as such was increasingly questioned in the final years of the nineteenth century:

> Although the dominant ethos until the 1890s was still the observationally oriented approach called positivism, with its literary correlate naturalism, a new attitude was visible on the horizon. The hegemony of what we have called Cartesian perspectivalism was beginning to unravel, leading initially to explorations of alternative scopic regimes (including those waiting to be recovered from earlier eras), and finally to a full-fledged critique of ocularcentrism in the twentieth century [...]. (Jay 1993, 146–147)

But not only scientific observation and the reliance on sight more broadly came to an end at the time when Doyle wrote his stories: these stories also overlap – and this is a second aspect – with the shift from the eighteenth- and nineteenth-century realist novel to the Modernist novel. Modernist literature – to risk some simplification here – moves away from the realist presumption of representing an exterior reality and is, instead, more interested in performing processes of consciousness than processes of vision.[1] What matters primarily is not how reality emerges in a specific visual process (namely, observation), but what charac-

---

1 For Modernism's rupture with the aesthetics of visualizing description, see also my remarks on Kafka in chapter 1. Kafka and Virginia Woolf (to whose work I will turn in a moment) offer certainly very different scenarios of the demise of visualizing description. Kafka fundamentally questions description's potential to render an image of exterior reality; for Woolf, by contrast, such exterior reality is no longer a primary concern.

https://doi.org/10.1515/9783110594348-175

ters think or feel about a given subject at a particular point in time.[2] This is not to say that Modernism is not interested in visual impulses, but what is more important than these impulses as such or the reality they reflect are the thought processes to which they give rise and through which these impulses are, in turn, interpreted. What I say here reflects a rather common understanding of Modernism, and so I will limit myself to just one example, a passage from the opening scene of Virginia Woolf's *To the Lighthouse* (1927), the very novel that already serves as the model of Modernism in Erich Auerbach's canonic study *Mimesis*.

In some sense, the following passage is highly visual. What is presented to us is the scene of a mother looking at her son, who is looking at pictures from an illustrated catalogue. But instead of dwelling on these sights themselves, the novel's main interest lies in the interpretation of these sights through the beholders:

> James Ramsay, sitting on the floor, cutting out pictures from the illustrated catalogue of the Army and Navy Stores, endowed the picture of a refrigerator as his mother spoke with heavenly bliss. It was fringed with joy. The wheelbarrow, the lawn-mower, the sound of poplar trees, leaves whitening before rain, rooks cawing, brooms knocking, dresses rustling—all these were so coloured and distinguished in his mind that he had already his private code, his secret language, though he appeared the image of stark and uncompromising severity, with his high forehead and his fierce blue eyes, impeccably candid and pure, frowning slightly at the sight of human frailty, so that his mother, watching him guide his scissors neatly round the refrigerator, imagined him all red and ermine on the Bench or directing a stern and momentous enterprise in some crisis of public affairs. (Woolf 2008, 7)

What is important in this passage are not the many visual perceptions (such as the image of the refrigerator), but the way in which these images are recast in the characters' minds (in the case of the refrigerator, "fringed with joy"). Even if there were a literary observation in this passage (instead of the mere collection of individual snapshots), the novel's main interest would likely not be in the re-

---

2 Erich Auerbach captures this shift in his analysis of Virginia Woolf's novel *To the Lighthouse* (1927): "Wenn es sich etwa um das Haus, oder um das Schweizer Dienstmädchen handelt, so wird uns nicht die objektive Kenntnis mitgeteilt, welche Virginia Woolf mit schöpferischer Einbildungskraft von diesen Gegenständen besitzt, sondern dasjenige, was Mrs. Ramsay in einem bestimmten Augenblick darüber denkt oder empfindet." (Auerbach 2015, 496) [Concerning the house or the Swiss maid, for instance, we are not given any objective information, which Virginia Woolf possesses through her creative imagination, but that which Mrs. Ramsay thinks or feels about these subjects at a specific moment.] For a recent critical revaluation of Auerbach's reading of *To the Lighthouse*, see Levenson 2015.

ality that appears in the observation, but in its interpretation by one of the characters.

But there is yet a third reason that allows me to identify in the Sherlock Holmes stories the historical end of the age of literary observation, and this is the fact that these stories fall in the very period in which the interest in *Le Diable boiteux* and its gesture to the visual suddenly vanishes: as far as I can see, Doyle's early Holmes story "A Case of Identity" is the last prominent fictional text that alludes to the flight over Madrid in Lesage's novel.[3] This abating interest in *Le Diable boiteux* confirms the sense that writers of the *fin-de-siècle* lost an interest in narratives that are grounded in the visual discovery of the world.

More importantly still, Doyle's stories provide a logical end to the theory of literary observation that I advance in this book. Collapsing into one the acts of description and narration, the seeing of a state and the recording of a sequence of events, the Sherlock Holmes stories present a powerful solution to the shortcomings that characterized the kinds of encounter with the visual that Lesage's and Rétif's novels exemplify. For Lesage, the presently visible and describable tableaux under the roofs of the houses of Madrid remain categorically separated from the devil's stories, which themselves cannot be seen. Rétif, by contrast, shows how the description of a striking image can be extended in the focused recording of a sequence of events to disclose a narrative. But although image and sequence are thus connected in the combined observational procedure of the nocturnal spectator, the transition from image to sequence remains a potentially precarious one – as my readings of Goethe, Büchner, and Poe have shown. In light of this long struggle with observational procedures in literature, the advantages of Holmes's detective gaze are at once obvious: the transition from image to sequence is located within the image itself. The right observation of the static image leads to the disclosure of a narrative sequence. But this perfect procedure of observation is, in Doyle's stories, bought at a high price. Holmes's detective work perpetually tips over into a classificatory calculus, and it thus tends to give up on observation's greatest asset: the openness to the "possibilities for new knowledge in the most unexpected places" (Daston, Lunbeck 2011, 8).

Characterized in this way, we arrive here at a neat closure – too neat to be true. What has to be taken into account is, first of all, the steady recurrence of literary observations in many works of the twentieth and twenty-first centuries.[4]

---

3 See my discussion of the reception of *Le Diable boiteux* in chapter 2.
4 The great legacy and lasting popularity of Sherlock Holmes is one obvious indicator of an afterlife of observation.

Describing a character (or object) and setting this character (or object) in motion: although I read this as an important reality effect, characteristic of the poetics (and period) of literary realism, this is too common a device to vanish completely at any particular point in history. And this is not all. We can also find still new and interesting variations of literary observations after 1900. Let me just gesture here to one especially curious example, taken from the beginning of the 1939 crime classic *The Big Sleep* by Raymond Chandler, a writer who quite consciously placed himself in the realist tradition.[5] In the first paragraph of this novel, we are introduced to the private detective Philipp Marlowe, and, at first sight, this paragraph follows, once more, the familiar structure of observation. Painted against a cursorily sketched background of the Los Angeles landscape appears the image of the detective, which is visualized through a detailed description of his clothing. In the paragraph's final sentence, we see this character turn to act, and thereby we also see the static image of description set in motion:

> It was about eleven o'clock in the morning, mid October, with the sun not shining and a look of hard wet rain in the clearness of the foothills. I was wearing my powder-blue suit, with dark blue shirt, tie and display handkerchief, black brogues, black wool socks with dark blue clocks on them. I was neat, clean, shaved and sober, and I didn't care who knew it. I was everything the well-dressed private detective ought to be. I was calling on four million dollars. (Chandler 2005, 1)

What is so peculiar about the highly visual opening of this novel – with all its specifications of the detective's dress – is that it overwhelms the reader with a strange confusion of perspective. On the one hand, the text performs in all clarity the mechanism of observation: it sketches the background of a gray, rainy day; it introduces against this background the detective, and with the last short sentence, we feel that we see him turn to an expensive big house (which is indeed being described in the subsequent paragraph) and thus set the plot of the novel in motion. And yet, the visual process that is triggered by the text happens almost *against* the text's own intention. The text does not simply perform vision through an observing character in the text (think of the schoolboy who serves as the observer-narrator in the opening scene of *Madame Bovary*). Nor is this text presented by a disembodied heterodiegetic narrator whose presentation of the story can imitate with the shift from description to narration the perceptual structure of observations (think of the very opening of *Le Diable boiteux*). Instead of being able to see the world through the eyes of the narrator or of a character in

---

5 On Raymond Chandler's ties to (nineteenth-century) realism in general and to the realism of the twentieth-century American novel more specifically, see Witschi 2002, esp. 139–166.

the text, we are visualizing this scene by looking at the very character who conveys the information – a character, moreover, who expresses a certain dislike of possible voyeurs: "I was neat, clean, shaved and sober, and I didn't care who knew it." It is not the case that the narrator would prohibit observation of himself, but he certainly rejects the notion that he is there to be looked at.

If it is not already evident that this novel, while citing the mechanisms of literary observations, also reflects on this very procedure in curious ways, let us take a look at the following paragraph, which contains a description of the building that the detective Philip Marlowe enters:

> The main hallway of the Sternwood place was two stories high. Over the entrance doors, which would have let in a troop of Indian elephants, there was a broad stained-glass panel showing a knight in dark armour rescuing a lady who was tied to a tree and didn't have any clothes on but some very long and convenient hair. The knight had pushed the vizor of his helmet back to be sociable, and he was fiddling with the knots on the ropes that tied the lady to the tree and not getting anywhere. I stood there and thought that if I lived in the house, I would sooner or later have to climb up there and help him. He didn't really seem to be trying. (Chandler 2005, 1)

In some sense, the text shifts with this second paragraph to a much more conventional perspective. Now we see the world of the novel through the eyes of the homodiegetic narrator, the detective Philip Marlowe. But that does not mean that the very conscious reflections on observational procedures have now been dropped. Quite to the contrary, they are at the center of this paragraph.

Marlowe focuses our attention on a glass panel depicting a knight, who, instead of freeing an undressed woman from a tree to which she is tied, just pushes his vizor up and seems content to be merely looking at her and to be "fiddling" with the ropes. In this account of the scene of the glass panel, at least two important aspects of observation are negotiated. First, this passage points us again to the voyeurism inherent in observation – a concern that shaped already the first paragraph. It might be worthwhile to unpack this aspect here a little. For while the scene on the glass panel appears to consist in a very simple constellation of male voyeurism, it is less clear how to read this scene in light of the novel's first paragraph. Crucially, while Marlowe imagines himself in the position of the male onlooker, whom he wants to assist, the reader will be just as likely to see Marlowe in the position of the naked woman tied to the tree. After all, in the first paragraph, Marlowe is the one who has exposed himself in all his outward appearance (including, of course, the "tie" that he, so to speak, shares with the tied woman) to the reader, and he is the novel's primary object of voyeurism. The opposition between observer and observed is called into question; the observer himself appears to be inadvertently the first object of his observations. Chandler's novel, in other words, insistently reflects in observational procedures the questions of who is getting to see, and who will be looked at.

However, beyond this question of voyeurism, this scene also contains a commentary on the agency of the observer in the shift from stillness to motion, from description of narration. Imagining entering the setting of his own description (the glass panel) as an active agent, the narrator literally sets the static scene of description in motion. Put differently, it is not the case that the world that the observer encounters changes itself from stillness to motion (from the object of description to the object of narration); instead, it is the observer who triggers this shift. Rather than just describing the setting of the following scenes (the image on the glass panel and the house into which he is about to enter), the text explicitly states through Marlowe's reflection on the glass panel that the observer himself (through his actions) turns the object of static description into the object of narration.[6] From the outset, *The Big Sleep* thus presents itself as a text that not only inherits realist literary procedures of observation, but that also wittily looks into the ethical and aesthetic implications of observation.

Needless to say, the opening of *The Big Sleep* stands here only as one example of how twentieth-century novelists arrive at interesting new variations and reflections on the literary procedure of observation. But if we want to study the productive legacy of literary observation in the twentieth and twenty-first centuries, we likely also have to look beyond literature. If we understand literary observation at its most basic as a reality effect created by the combination of the description of images with the narration of sequences in which these images are set in motion, then we do find something like this also in other media. We just have to think back here once more to the metaphor used by Jean Varloot to describe the nocturnal spectator of Rétif's novel: the nocturnal spectator, Varloot says, "cerne un image et filme" (Varloot 1987, 12) [sketches an images and films]. This phrasing is suggestive of the idea that perhaps the most important continuation of literary observation outside literature occurs in the medium of film. In closing, I will highlight two filmic examples that contain a rather explicit reflection on film's ability to produce observations.

In the chapter on Sherlock Holmes, I already gestured toward Antonioni's *Blow Up* as an example of a film that shows how narrative can be interrupted to focus on an image that is, in turn, extended into a sequence. As Antonioni's photographer produces increasingly blown-up versions of a single photograph,

---

6 However, the confusion about the role that Marlowe represents in the image on the glass panel raises again some questions about the turn the narrative will take. Most importantly, we do not know whose image, so to speak, will be set in motion (that is, who will change) – that of Marlowe's environment or that of Marlowe himself.

he reconstructs the narrative of a murder captured in this photo. But in showing the series of blown-up prints hanging next to each other in the photographer's studio, Antonioni also cunningly transforms the medium of photography into that of film (which consists precisely in a series of photos shown quickly one after the other). *Blow Up*, in other words, offers us an allegory of film's fundamental ability to set static images in motion and to produce reality effects.

I just want to add here one further filmic example, equally famous, but somewhat more emphatic or dramatic in the way in which it performs the moment in which film sets static images in motion. This example is taken from *The Empire Strikes Back* (1980, directed by Irvin Kershner), the second part of the original *Star Wars* trilogy. At one point in this film, Han Solo and his crew on the Millennium Falcon seek shelter in what they take to be a cave on an asteroid. However, as they leave the ship and explore their surroundings, the moist environment in which they find themselves begins to seem increasingly strange. In a final test, Han Solo shoots his gun at the ground, and, as a result, the ground starts shaking. Han Solo now realizes that they have, in fact, landed inside the gullet of a space monster. This striking scene figures prominently in the opening of Amitav Ghosh's recent essay *The Great Derangement: Climate Change and the Unthinkable* (Ghosh 2016, 3). Ghosh reads this episode as a scene of recognition in which humanity realizes that its environment is alive and, on some level, an equal agent to be reckoned with. In view of the theory of observation advanced in this book, however, the scene from *The Empire Strikes Back* seems to be interested not only in the binary of dead and living matter, but also in the transition from static to moving pictures. The moment in which we see the earth shake and see Han Solo lift his arms in the air, trying to keep his balance, stages film's fundamental possibility to set static pictures in motion and thereby to show reality in its full dynamic visuality.

As the scenes from *The Big Sleep*, *Blow Up*, and *The Empire Strikes Back* indicate, literary observation has a robust afterlife beyond the historical development of the realist novel in the eighteenth and nineteenth centuries and beyond the time in which 'observation' reigned supreme in the sciences. In some ways, however, we were well prepared for this insight from the various readings in this book. For while I discussed literary observation largely as a procedure of the period of realism (broadly conceived) that reflects certain elements of the discourse on observation in the natural sciences (notably concerning the relation between image and sequence), it also became clear that the stylistics of literary observation intersect with a wide range of other discourse formations from art history through psychology to urban surveillance. And it would be an illusion to think either that all these various discourses vanished around 1900, or that whatever I have analyzed in this book could offer a complete account of the cultural contexts that intersect with the literary procedure of observation. In a similar way, then, in which the reality effect of observation (in which a static image appears and is subsequently set in motion) may serve as a possible paradigm of what literary realism consists in more broadly (visualizing a dynamic world and mimicking processes of perception), the focus on the eighteenth and nineteenth centuries offered us some paradigmatic constellations of how literary observations can function – or fail to function – in specific cultural and aesthetic constellations more generally.

Bibliography

Adler, Judith. "Origins of Sightseeing." *Annals of Tourism Research* 16 (1989): 7–29.

Alpers, Svetlana. "Describe or Narrate? A Problem in Realistic Representation." *New Literary History* 8.1 (1976): 15–41.

Amaral, Genevieve. "Edgar Allan Poe's Fear of Texts: *The Man of the Crowd* as Literary Monster." *The Comparatist* 35.1 (2011): 227–238.

Arac, Jonathan. *Commissioned Spirits: The Shaping of Social Motion in Dickens, Carlyle, Melville, and Hawthorne.* New Brunswick: Rutgers University Press, 1979.

Assel, Jutta: "Werther-Illustrationen. Bilddokumente als Rezeptionsgeschichte." http://www. goethezeitportal.de/wissen/illustrationen/johann-wolfgang-von-goethe/die-leiden-des-jungen-werther/jutta-assel-werther-illustrationen-bilddokumente-als-re zeptionsgeschichte.html *Goethezeitportal* 2002–2018 (10 January 2018).

Auerbach, Erich. *Mimesis: Dargestellte Wirklichkeit in der abendländischen Kultur.* 11th edition. Tübingen: Narr Francke Atempto, 2015.

Auerbach, Jonathan. *The Romance of Failure: First-Person Fictions of Poe, Hawthorne, and James.* New York: Oxford University Press, 1989.

Bacon, Francis. *Novum Organum.* Ed. Thomas Fowler. Oxford: Clarendon Press, 1889.

Bacon, Francis. *The New Organon.* Ed. Lisa Jardine and Michael Silverthorne. Trans. Michael Silverthorne. New York: Cambridge University Press, 2000.

Barloon, Jim. "The Case for Identity: Sherlock Holmes and the Singular Find." *Clues* 25.1 (2006): 33–44.

Barr, Philippe. *Rétif de la Bretonne spectateur nocturne: Une esthétique de la pauvreté.* Amsterdam: Rodopi, 2012.

Barthes, Roland. "L'effet de réel." *Communications* 11 (1968): 84–89.

Barthes, Roland. *Camera Lucida: Reflections on Photography.* Trans. Richard Howard. New York: Hill and Wang, 1981.

Barthes, Roland. "The Reality Effect." In *The Rustle of Language.* Trans. Richard Howard. Oxford: Basil Blackwell, 1986. 141–148.

Baudelaire, Charles. *Le Peintre de la vie moderne.* Ed. Silvia Acierno and Julio Baquero Cruz. Paris: Sandre, 2009.

Behn, Aphra. *Oroonoko, The Rover and Other Works.* Ed. Janet Todd. London: Penguin, 1992.

Bender, John, and Michael Marrinan, eds. *Regimes of Description: In the Archive of the Eighteenth Century.* Stanford: Stanford University Press, 2005.

Benjamin, Walter. *Illuminationen: Ausgewählte Schriften 1.* Frankfurt am Main: Suhrkamp, 1961.

Benjamin, Walter. "Kleine Geschichte der Photographie." In *Gesammelte Schriften* 2.1. Ed. Rolf Tiedemann and Hermann Schweppenhäuser. Frankfurt am Main: Suhrkamp, 1991. 368–385.

Benjamin, Walter. "Paris: Capital of the Nineteenth Century." *Perspecta* 12 (1969): 163–172.

Beaujour, Michel. "Some Paradoxes of Description." *Yale French Studies* 61 (1981): 27–59.

Bobis, Laurence, and James Lequeux. "Cassini, Rømer and the Velocity of Light." *Journal of Astronomical History and Heritage* 11.2 (2008): 97–195.

Borgards, Roland. "Lenz." In *Büchner Handbuch.* Ed. Roland Borgards and Harald Neumeyer. Stuttgart: Metzler, 2009. 51–70.

Bortoft, Henri. *The Wholeness of Nature: Goethe's Way toward a Science of Conscious Participation in Nature.* Hudson: Lindisfarne, 1996.

Botero, Giovanni. *Allgemeine Weltbeschreibung.* Köln: Johann Gymnich Erben, 1596.

Breton, André. *Manifestoes of Surrealism.* Trans. Richard Seaver and Helen R. Lane. Ann Arbor: The University of Michigan Press, 1969.

Brooks, Peter. *Realist Vision.* New Haven: Yale University Press, 2005.

Büchner, Georg. *Dichtungen.* Ed. Henri Poschmann and Rosemarie Poschmann. Frankfurt am Main: Deutscher Klassiker Verlag im Taschenbuch, 2006.

Büchner, Georg. *The Major Works.* Ed. Matthew Wilson Smith. Trans. Henry J. Schmidt. New York: Norton, 2012.

Burton, Richard D. E. *The Flaneur and His City.* Durham: University of Durham, 1994.

Campe, Rüdiger. *Spiel der Wahrscheinlichkeit: Literatur und Berechnung zwischen Pascal und Kleist.* Göttingen: Wallstein, 2002.

Campe, Rüdiger. *The Game of Probability: Literature and Calculation from Pascal to Kleist.* Trans. Ellwood J. Wiggins Jr. Stanford: Stanford University Press, 2012.

Campe, Rüdiger, Jocelyn Holland, and Elisabeth Strowick. "Observation in Science and Literature: Preface." *Monatshefte* 105.3 (2013): 371–375.

Chandler, Raymond. *The Big Sleep.* Introd. Ian Rankin. London: Penguin, 2005.

Cohen, Bernhard. "Roemer and the First Determination of the Velocity of Light." *Isis* 31.1 (1940): 327–379.

Crary, Jonathan. *Techniques of the Observer: On Vision and Modernity in the Nineteenth Century.* Cambridge, Massachusetts: MIT Press, 1990.

Crary, Jonathan. *Suspensions of Perception: Attention, Spectacle, and Modern Culture.* Cambridge, Massachusetts: The MIT Press, 1999.

Crews, Frederick. *The Memory Wars: Freud's Legacy in Dispute.* New York: New York Review of Books, 1995.

Culler, Jonathan. "The Realism of Madame Bovary." *MLN* 122.4 (2007): 683–696.

Darwin, Charles. *The Voyage of the Beagle.* Ware: Wordsworth, 1997.

Darwin, Charles. *On the Origin of Species.* Ed., introd. Gillian Beer. Oxford: Oxford University Press, 2008.

Daston, Lorraine. *Eine kurze Geschichte der Aufmerksamkeit.* München: Carl Friedrich von Siemens Stiftung, 2000.

Daston, Lorraine. "The Disciplines of Attention." In *A New History of German Literature.* Ed David Wellbery. Cambridge, Massachusetts: Harvard University Press, 2004. 434–440.

Daston, Lorraine. "Description by Omission: Nature Enlightened and Obscured." In *Regimes of Description: In the Archive of the Eighteenth Century.* Ed. John Bender and Michael Marrinan. Stanford: Stanford University Press, 2005. 11–24.

Daston, Lorraine. "On Scientific Observation." *Isis* 99.1 (2008): 97–110.

Daston, Lorraine. "The Empire of Observation 1600–1800." In *Histories of Scientific Observation.* Ed. Lorraine Daston and Elizabeth Lunbeck. Chicago: University of Chicago Press, 2011. 81–113.

Daston, Lorraine, and Peter Galison. *Objectivity.* New York: Zone Books, 2007.

Daston, Lorraine, and Elizabeth Lunbeck, eds. *Histories of Scientific Observation.* Chicago: University of Chicago Press, 2011.

Delon, Michel. "Le détail, le réel et le réalisme dans la perspective française." *Studies on Voltaire and the Eighteenth Century* 2 (2009): 15–28.

Dotzler, Bernhard. "Werthers Leser: Über die Appellstruktur der Texte im Licht von Goethes
  Romanen." *Modern Language Notes* 114.3 (1999): 445–470.
Doyle, Arthur Conan. *The Complete Sherlock Holmes*. Preface by Christopher Morley. Garden
  City: Doubleday, 1953.
Drügh, Heinz. *Ästhetik der Beschreibung: Poetische und kulturelle Energie deskriptiver Texte
  (1700–2000)*. Tübingen: Francke, 2006.
Eco, Umberto. "Horns, Hooves, Insteps: Some Hypotheses on Three Types of Abduction." In
  The Sign of Three: Dupin, Holmes, Peirce. Ed. Umberto Eco, Thomas A. Sebeok, and Jean
  Umiker-Sebeok. Bloomington: Indiana University Press, 1983. 198–220.
Edmiston, William F. "Public Protection or Social Repression? Restif de la Bretonne and the
  Role of the State." *South Atlantic Review* 59.1 (1994): 45–64.
Fick, Monika. *Lessing-Handbuch: Leben, Werk, Wirkung*. Stuttgart: Metzler, 2000.
Fischer, Luke. "Goethe contra Hegel: The Question of the End of Art." *Goethe Yearbook* 18
  (2011): 127–157.
Flaschka, Horst. *Goethes 'Werther': Werkkontextuelle Deskription und Analyse*. München: Fink,
  1987.
Flaubert, Gustave. *L'Éducation sentimentale: Histoire d'un jeune homme*. Introd. Pierre
  Sipriot. Paris: Le Livre de Poche, 1983.
Flaubert, Gustave. *Madame Bovary: Moeurs de province*. Paris: Gallimard, 1998.
Flaubert, Gustave. *Madame Bovary: Provincial Manners*. Trans. Margaret Mauldon. Introd.
  Malcolm Bowie. Notes Mark Overstall. Oxford: Oxford Univesity Press, 2004. [=Flaubert
  2004a]
Flaubert, Gustave. *Sentimental Education*. Trans. Robert Baldick. Introd. Geoffrey Wall.
  London: Penguin, 2004. [=Flaubert 2004b]
Fleming, David. *City of Rhetoric: Revitalizing the Public Sphere in Metropolitan America*.
  Albany: State University of New York Press, 2008.
Förster, Eckart. "Goethe and the 'Auge des Geistes.'" *Deutsche Vierteljahresschrift für
  Literaturwissenschaft und Geistesgeschichte* 75.1 (2001): 87–101.
Förster, Eckart. *The Twenty-Five Years of Philosophy: A Systematic Reconstruction*. Trans.
  Brady Bowman. Cambridge, Massachusetts: Harvard University Press, 2012.
Foucault, Michel. *Surveiller et punir: naissance de la prison*. Paris: Gallimard, 1975.
Foucault, Michel. *Discipline and Punish: The Birth of the Prison*. Trans. Alan Sheridan. New
  York: Vintage Books, 1977.
Foucault, Michel. *The History of Sexuality*. Volume 1. Trans. Robert Hurley. New York:
  Pantheon Books, 1978.
Frank, Lawrence. *Victorian Detective Fiction and the Nature of Evidence: The Scientific
  Investigations of Poe, Dickens, and Doyle*. Houndmills: Palgrave MacMillan, 2003.
Freud, Sigmund. *The Complete Edition of the Psychological Works of Sigmund Freud*. Ed.,
  trans. James Strachey. London: The Hogarth Press, 1955.
Freud, Sigmund. *Studienausgabe*. Ed. Alexander Mitscherlich. Frankfurt am Main: Fischer,
  2000.
Fried, Michael. *Absorption and Theatricality: Painting and Beholder in the Age of Diderot*.
  Chicago: University of Chicago Press, 1988.
Genette, Gérard. *Figures of Literary Discourse*. Trans. Alan Sheridan. Introd. Marie-Rose
  Logan. New York: Columbia University Press, 1982.

Geulen, Eva. "Depicting Description: Lukács and Stifter." *The Germanic Review* 73.3 (1998): 267–279.

Ghosh, Amitav. *The Great Derangement: Climate Change and the Unthinkable.* Chicago: University of Chicago Press, 2016.

Ginzburg, Carlo. "Morelli, Freud and Sherlock Holmes: Clues and Scientific Method." Trans. Anna Davin. *History Workshop* 9 (1980): 5–36.

Goethe, Johann Wolfgang. *Novels and Tales.* Trans. R.D. Boylan. London: Henry G. Bohn, 1854.

Goethe, Johann Wolfgang. *Faust, The Sorrows of Young Werther.* Trans. James Stuart Blackie. Oxford: Oxford University Press, 1970.

Goethe, Johann Wolfgang. *Werke. Naturwissenschaftliche Schriften 1.* Ed. Dorothea Kuhn and Rilke Wankmüller. München: C.H. Beck, 1981.

Goethe, Johann Wolfgang. *Aus meinem Leben: Dichtung und Wahrheit.* Ed. Klaus-Detlef Müller. Frankfurt am Main: Deutscher Klassiker Verlag, 1986.

Goethe, Johann Wolfgang. *Dramen 1776–1790.* Ed. Dieter Borchmeyer. Frankfurt am Main: Deutscher Klassiker Verlag, 1988.

Goethe, Johann Wolfgang. *Campagne in Frankreich, Belagerung von Mainz, Reiseschriften.* Ed. Klaus-Detlef Müller. Frankfurt am Main: Deutscher Klassiker Verlag, 1994.

Goethe, Johann Wolfgang. *Scientific Studies.* Ed., trans. Douglas Miller. Princeton: Princeton University Press, 1995.

Goethe, Johann Wolfgang. *Faust.* Ed. Albrecht Schöne. Frankfurt am Main: Deutscher Klassiker Verlag im Taschenbuch, 2005.

Goethe, Johann Wolfgang. *Die Leiden des jungen Werthers, Die Wahlverwandtschaften, Kleine Prosa, Epen.* Ed. Christoph Brecht and Waltraud Wiethölter. Frankfurt am Main: Deutscher Klassiker Verlag im Taschenbuch, 2006.

Goulemot, Jean-Marie. "Effets de réalité et constructions narratives dans les *Les Nuits Revolutionnaire.*" *Études Rétiviennes* 11 (1989): 201–215.

Grimm, Jacob, and Wilhelm Grimm (ed.). *Deutsches Wörterbuch.* Leipzig: S. Hirzel, 1854–1961.

Harron, Philippe. *Introduction à l'analyse du descriptif.* Paris: Hachette, 1981.

Harron, Philippe. *La Description littéraire: De l'Antiquité à Roland Barthes: une anthologie.* Paris: Macula, 1991.

Hayes, Kevin J. "Retzsch's Outlines and Poe's 'The Man of the Crowd.'" *Gothic Studies* 12.2 (2010): 29–41.

Heller, Otto. Preface. In *Das Haidedorf.* By Adalbert Stifter. Ed. Otto Heller. Boston: D.C. Heath & Co., 1891. iii–iv.

Hoffmann, E.T.A. *Sämtliche Werke in sechs Bänden.* Ed. Hartmut Steinecke and Wulf Segebrecht. Frankfurt am Main: Deutscher Klassiker Verlag, 1985–2004.

Hofmann, Christoph. *Unter Beobachtung: Naturforschung in der Zeit der Sinnesapparate.* Göttingen: Wallstein, 2006.

Holmes, Richard. *The Age of Wonder: How the Romantic Generation Discovered the Beauty and Terror of Science.* London: Harper Press, 2008.

Holmes, Tove. "'…Was ich in diesem Hause geworden bin.': Adalbert Stifter's Visual Curriculum." *Zeitschrift für deutsche Philologie* 129.4 (2010): 559–577.

Holub, Robert C. "The Paradoxes of Realism: An Examination of the *Kunstgespräch* in Büchner's *Lenz.*" *Deutsche Vierteljahresschrift für Literaturwissenschaft und Geistesgeschichte* 59.1 (1985): 102–124.

Horace. *Satires, Epistles and Ars Poetica.* Trans. H. Rushton Fairclough. Cambridge, Massachusetts: Harvard University Press, 1942.

Huber, Martin. *Der Text als Bühne: Theatrales Erzählen um 1800.* Göttingen: Vandenhoeck & Ruprecht, 2002.

Iknayan, Marguerite. "The Fortunes of 'Gil Blas' in the Romantic Period." *The French Review* 31.5 (1958): 370–377.

Innocenti, Beth. "Towards a Theory of Vivid Description in Cicero's 'Verrine' Orations." *Rhetorica* 14.4 (1994): 355–381.

Jacob, P.L. *Bibliographie et Iconographie des tous les ouvrages de Restif de la Bretonne.* Paris: Auguste Fontaine, 1875.

Jacobson, Roman. "On Realism in Art." In *Language in Literature.* Ed. Krystina Pomorska and Stephen Rudy. Cambridge: The Belknap Press of Harvard University Press, 1987. 19–27.

James, Louis. *The Victorian Novel.* Oxford: Blackwell, 2006.

Jameson, Frederic. *Signatures of the Visible.* New York: Routledge, 1992.

Jansen, Peter K. "The Structural Function of the *Kunstgespräch* in Büchner's *Lenz.*" *Monatshefte für deutschsprachige Literatur und Kultur* 67.2 (1975): 145–156.

Jaucourt, Louis de. "Spectacles." In *Encyclopédie ou Dictionnaire raisonné des sciences, des arts et des métiers, etc.* Ed. Denis Diderot and Jean le Rond d'Alembert. Paris: 1751–1780. Vol. 15, 446–447.

Jay, Martin. *Downcast Eyes: The Denigration of Vision in Twentieth-Century French Thought.* Berkeley: University of California Press, 1993.

Kablitz, Andreas. "Realism as a Poetics of Observation." In *What Is Narratology?: Questions and Answers Regarding the Status of a Theory.* Ed. Fotis Jannidis, John Pier, and Wolf Schmid. Berlin: De Gruyter, 2003. 99–135.

Kafka, Franz. *Briefe 1902–1924.* Frankfurt am Main: S. Fischer, 1966.

Kafka, Franz. *Collected Stories.* Trans. Willa and Edwin Muir. Ed. Gabriel Josipovici. New York: Everyman's Library, 1993.

Kafka, Franz. *Das Schloß.* Frankfurt am Main: S. Fischer, 1994.

Kafka, Franz. *The Castle.* Trans. Mark Harman. New York: Schocken Books, 1998.

Kafka, Franz. *Die Erzählungen und andere ausgewählte Prosa.* Frankfurt am Main: S. Fischer, 2002.

Kasson, John F. *Rudeness and Civility: Manners in Nineteenth-Century Urban America.* New York: Hill and Wang, 1990.

Kennedy, J. Gerald. "The Limits of Reason: Poe's Deluded Detectives." *American Literature* 47.2 (1975): 184–196.

Kennedy, J. Gerald. Rev. of *The Sign of Three: Dupin, Holmes, Peirce,* by Umberto Eco, Thomas A. Sebeok, and Jean Umiker-Sebeok. *Philosophy and Literature* 10.1 (1986): 122–123.

Kerr, Douglas. *Conan Doyle: Writing, Profession, and Practice.* Oxford: Oxford University Press, 2013.

Klausnitzer, Ralf. "Beobachten." In *Literatur und Wissen. Ein interdisziplinäres Handbuch.* Ed. Roland Borgards, Harald Neumeyer, Nicolas Pethes, and Yvonne Wübben. Stuttgart: Metzler, 2013. 241–253.

Klein, Claude. "La narration et l'imaginaire dans *Les Nuits de Paris.*" *Études Rétiviennes* 21 (1994): 171–184.

Klotz, Volker. *Die erzählte Stadt: Ein Sujet als Herausforderung des Romans von Lesage bis Döblin.* München: Hanser, 1969.

Koselleck, Reinhart. *Futures Past: On the Semantics of Historical Time.* Trans. Keith Tribe. New York: Columbia University Press, 2004.

Kuhn, Dorothea. "Ihr naht euch wieder, schwankende Gestalten." *Jahrbuch der Goethe-Gesellschaft: Neue Folge* 14/15 (1952/1953): 347–349.

La Bruyère, Jean de. *Les caractères de Theophraste traduits du grec avec Les caractères ou les moeurs de ce siècle.* Ed. Robert Garapon. Paris: Garnier, 1962.

Laufer, Roger. Préface. *Le Diable boiteux.* By Alain-René Lesage. Paris: Éditions Gallimard, 1984. 7–24.

Lauster, Martina. "Walter Benjamin's Myth of the Flâneur." *The Modern Language Review* 102.1 (2007): 139–156.

Le Sage, Alain-René. *Le Diable boiteux.* Paris: Veuve Pierre Ribou, 1726.

Le Sage, Alain-René. *Histoire de Gil Blas de Santillane.* Ed. Étiemble. Paris: Gallimard, 1973.

Le Sage, Alain-René. *Le Diable boiteux.* Ed. Roger Laufer. Paris: Éditions Gallimard, 1984.

Lessing, Gotthold Ephraim. "Laokoon." In *Werke in sechs Bänden.* Ed. Fritz Fischer. Zürich: Stauffacher, 1965. Vol. 5, 7–180.

Lessing, Gotthold Ephraim. *Laocoön: An Essay on the Limits of Poetry and Painting.* Trans. Edward Allen McCormick. Baltimore: Johns Hopkins University Press, 1984.

Levenson, Michael. "Narrative Perspective in *To the Lighthouse.*" In *The Cambridge Companion to* To the Lighthouse. Ed. Allison Pease. New York: Cambridge University Press, 2015. 19–29.

Luhmann, Niklas. *Die Kunst der Gesellschaft.* Frankfurt am Main: Suhrkamp, 1997.

Lukács, Georg. *Probleme des Realismus.* Berlin: Aufbau Verlag, 1955.

Lukács, Georg. *Writer and Critic and Other Essays.* Ed., trans. Arthur Kahn. London: Merlin Press, 1970.

Lyell, Charles. *Principles of Geology.* Ed., introd. James A. Secord. London: Penguin, 1997.

Mall, Laurence. Review of *Rétif de la Bretonne spectateur nocturne: Une esthétique de la pauvreté,* by Philippe Barr. *French Studies* 67.3 (2013): 412.

Marlowe, Christopher. *Doctor Faustus.* Ed. Sylvan Barnet. New York: Signet Classics, 1969.

Mayer, Wilhelm. "Zum Problem des Dichters Lenz." *Archiv für Psychiatrie und Nervenkrankheiten* 62.3 (1921): 889–890.

Meglin, Joellen A. "*Le Diable boiteux.* French Society Behind a Spanish Façade." *Dance Chronicle* 17.3 (1994): 263–302.

Menuret, Jean-Joseph. "Observateur." *Encyclopédie ou Dictionnaire raisonné des sciences, des arts et des métiers, etc.* Ed. Denis Diderot and Jean le Rond d'Alembert. Paris: 1751–1780. Vol. 11, 310–313.

Moretti, Franco. *The Way of the World: The Bildungsroman in European Culture.* London: Verso, 1987.

Müller-Salget, Klaus. "Zur Struktur von Goethes *Werther.*" In *Literatur ist Widerstand: Aufsätze aus drei Jahrzehnten.* Innsbruck: Universität Innsbruck, 2005. 73–87.

Nisle, Julius. *Göthe-Gallerie: Stahlstiche zu Göthes Meisterwerken nach Zeichnungen von Julius Nisle.* Stuttgart: Literatur-Comptoir, 1840.

North, Paul. *The Problem of Distraction.* Stanford: Stanford University Press, 2011.

Ogilvie, Brian W. *The Science of Describing: Natural History in Renaissance Europe*. Chicago: University of Cicago Press, 2006.

Parker, John J. "Some Remarks on Büchner's *Lenz* and Its Principal Source, The Oberlin Record." *German Life and Letters* 21.2 (1968): 103–111.

Peacham, Henry. *The Garden of Eloquence* (1577). Reprint. Menston: The Scolar Press Limited, 1971.

Pethes, Nicolas. "'Er ist ein interessanter casus, Subjekt Woyzeck': Büchner's Fallgeschichten." *Amsterdamer Beiträge zur neueren Germanistik* 81.1 (2012): 211–229.

Phillips, Natalie M. *Distraction: Problems of Attention in Eighteenth-Century Literature*. Baltimore: John Hopkins University Press, 2016.

Piper, Andrew. "Mapping Vision: Goethe, Cartography, and the Novel." In *Spatial Turns: Space, Place, and Mobility in German Literary and Visual Culture*. Ed. Jaimey Fisher and Barbara Mennel. Amsterdam: Rodopi, 2010. 27–51.

Poe, Edgar Allan. *The Complete Works*. Ed. James A. Harrison. New York: AMS Press, 1965.

Puttenham, George. *The Arte of English Poesie*. Ed. Edward Arber. London 1869.

Quintilian. *Institutio Oratoria*. Trans. H. E. Butler. Cambridge, Massachusetts: Harvard University Press, 1976.

Rachman, Stephen. "Reading Cities: Devotional Seeing in the Nineteenth Century." *American Literary History* 9.4 (1997): 653–675.

Reiffers, Moritz. *Das Ganze im Blick: Eine Kulturgeschichte des Überblicks vom Mittelalter bis zur Moderne*. Bielefeld: transcript, 2013.

[Rétif de la Bretonne, Nicolas-Edme.] *Les Nuits de Paris, ou le Spectateur-nocturne*. London [Paris]: 1788.

Rétif de la Breton[n]e, [Nicolas-Edme.] *Les nuits de Paris, ou l'observateur nocturne*. London [Paris]: 1789.

Restif de la Bretonne, Nicolas-Edme. *Les Nuits de Paris; Or, The Nocturnal Spectator: A Selection*. Trans. Linda Asher and Ellen Fertig. Introd. Jacques Barzun. New York: Random House, 1964.

Restif de La Bretonne, Nicolas-Edme, and Louis Sébastien Mercier. *Paris le jour, Paris la nuit: Tableau de Paris, Le nouveau Paris, Les Nuits de Paris*. Ed. Michel Delon. Paris: Robert Laffont, 1990.

Riha, Karl. "Aëro-Großstadt-Satire: Zu Lesages *Hinkendem Teufel*." In *Kritik, Satire, Parodie: Gesammelte Aufsätze*. Opladen: Westdeutscher Verlag, 1992, 17–26.

Ronen, Ruth. "Description, Narrative and Representation." *Narrative* 5.3 (1997): 274–286.

Rudwick, Martin J. S. *Worlds before Adam: The Reconstruction of Geohistory in the Age of Reform*. Chicago: University of Chicago Press, 2010.

Saint-Amour, Paul K. "The Vertical Flaneur: Narrational Tradecraft in the Colonial Metropolis." In *Joyce, Benjamin and Magical Urbanism*. Ed. Maurizia Boscagli and Edda Duffy. Amsterdam: Rodopi, 2011.

Sakellariadis, Spyros. "Descartes' Experimental Proof of the Infinite Velocity of Light." *Archive for History of Exact Sciences* 26.1 (1982): 1–12.

Sauerlandt, Max. *Michelangelo*. Düsseldorf: Karl Robert Langewiesche, 1911.

Scherpe, Klaus R. "Beschreiben, nicht Erzählen!: Beispiele zu einer ästhetischen Opposition." *Zeitschrift für Germanistik* 2 (1996): 368–383.

Schmid, Wolf. "Narrativity and Eventfulness." In *What Is Narratology?: Questions and Answers Regarding the Status of a Theory*. Ed. Fotis Jannidis, John Pier, and Wolf Schmid. Berlin: De Gruyter, 2003. 17–33.

Schmid, Wolf. *Narratology: An Introduction*. Trans. Alexander Starritt. Berlin: De Gruyter, 2010.

Schnyder, Peter. "Schrift, Bild, Sammlung, Karte: Medien geologischen Wissens in Stifters *Nachsommer*." In *Figuren der Übertragung: Adalbert Stifter und das Wissen seiner Zeit*. Ed. Michael Gamper and Karl Wagner. Zürich: Chronos-Verlag, 2009. 235–248.

Schöne, Albrecht. *Interpretationen zur dichterischen Gestaltung des Wahnsinns*. Diss. Münster: 1952.

Sebeok, Thomas A., and Jean Umiker-Sebeok. "'You Know My Method': A Juxtaposition of Charles S. Peirce and Sherlock Holmes." *Semiotica* 26.3–4 (1979): 203–250.

Senrett, Richard. *The Fall of Public Man: On the Social Psychology of Capitalism*. Cambridge: Cambridge University Press, 1977.

Shapin, Steven, and Simon Schaffer. *Leviathan and the Air-Pump: Hobbes, Boyle, and Experimental Life*. Princeton: Princeton University Press, 1985.

Shapple Spillman, Deborah. *British Colonial Realism in Africa: Inalienable Objects, Contested Domains*. Houndmills: Palgrave Macmillan, 2012.

Stalnaker, Joanna. *Unfinished Enlightenment: Description in the Age of the Encyclopedia*. Ithaca: Cornell University Press, 2010.

Stern, J.P. *Re-Interpretations: Seven Studies in Nineteenth-Century German Literature*. London: Thames and Hudson, 1964.

Stifter, Adalbert. "Bergkristall." In *Gesammelte Werke in vierzehn Bänden*. Ed. Konrad Steffen. Basel: Birkhäuser Verlag, 1962–1972. Vol. 4. 181–241.

Stifter, Adalbert. *Rock Crystal*. Trans. Elizabeth Mayer and Marianne Moore. Introd. W.H. Auden. New York: New York Review of Books, 2008.

Sweeney, Susan Elizabeth. "The Magnifying Glass: Spectacular Distance in Poe's 'Man of the Crowd' and Beyond." *Poe Studies/Dark Romanticism* 36.1–2 (2003): 3–17.

Turcot, Laurent. "Du promeneur au flâneur: les influences de Rétif dans la construction d'une figure sociale du XIXe au XXIe siècle." *Etudes rétiviennes* 38 (2006): 143–156.

Turcot, Laurent. "Promenades et flâneries à Paris du XVIIe au XXIe siècles: la marche comme construction d'une identité urbaine." In *Marcher en ville: Faire corps, prendre corps, donner corps aux ambiances urbaines*. Paris: Éditions des archives contemporaines, 2010. 65–84.

Varloot, Jean. "Préface." In *Les Nuits de Paris ou le Spectateur-nocturne*. By Nicolas-Edme Rétif de la Bretonne. Ed. Michel Delon. 2nd edition. Paris: Gallimard, 1987. 7–27.

Vic, Jean. "La composition et les sources du *Diable boiteux* de Lesage." *Revue d'histoire littéraire de la France* 27.4 (1920): 481–517.

Vogl, Joseph. "Wolkenbotschaft." *Archiv für Mediengeschichte* 5 (2005): 69–79.

Wagner, Martin. "'… so ganz in dem Gefühl vom ruhigen Daseyn versunken, daß meine Kunst darunter leidet …' Michael Frieds *Absorption and Theatricality* und Goethes *Die Leiden des jungen Werthers*." *Jahrbuch der Grillparzer Gesellschaft* 3.24 (2011–2012): 183–226.

Wagner, Martin. "Sherlock Holmes and the Fiction of Agency." In *Sherlock Holmes in Context*. Ed. Sam Naidu. London: Palgrave MacMillan, 2017. 133–147.

Wall, Cynthia Sundberg. *The Prose of Things: Transformations of Description in the Eighteenth Century.* Chicago: University of Chicago Press, 2006.

Watt, Ian P. *The Rise of the Novel: Studies in Defoe, Richardson and Fielding.* London: Chatto and Windus, 1957.

Wellbery, Caroline. "Mirrors to Images: The Transformation of Sentimental Paradigms in Goethe's *The Sorrows of Young Werther.*" *Studies in Romanticism* 25.2 (1986): 231–249.

Wellbery, David E. *Lessing's Laocoon: Semiotics and Aesthetics in the Age of Reason.* Cambridge: Cambridge University Press, 1984.

Wieland, Christoph Martin. *Sämmtliche Werke.* Leipzig: Verlag von Georg Joachim Göschen, 1839.

Williams, Raymond. *Keywords: A Vocabulary of Culture and Society.* New York: Oxford University Press, 1976.

Witschi, Nicolaus S. *Traces of Gold: California's Natural Resources and the Claim to Realism in Western American Literature.* Tuscaloosa: University of Alabama Press, 2002.

Woolf, Virginia. *To the Lighthouse.* Ed. David Bradshaw. Oxford: Oxford University Press, 2008.

Wyngaard, Amy S. "New Perspectives on Rétif de la Bretonne. Introduction." *Symposium* 60.3 (2006): 131–133.

Wyngaard, Amy S. *Bad Books: Rétif de la Bretonne, Sexuality, and Pornography.* Newark: Rowman & Littlefield, 2012.

Index

https://doi.org/10.1515/9783110594348-192